WILD FOOD PLANTS OF THE SIERRA

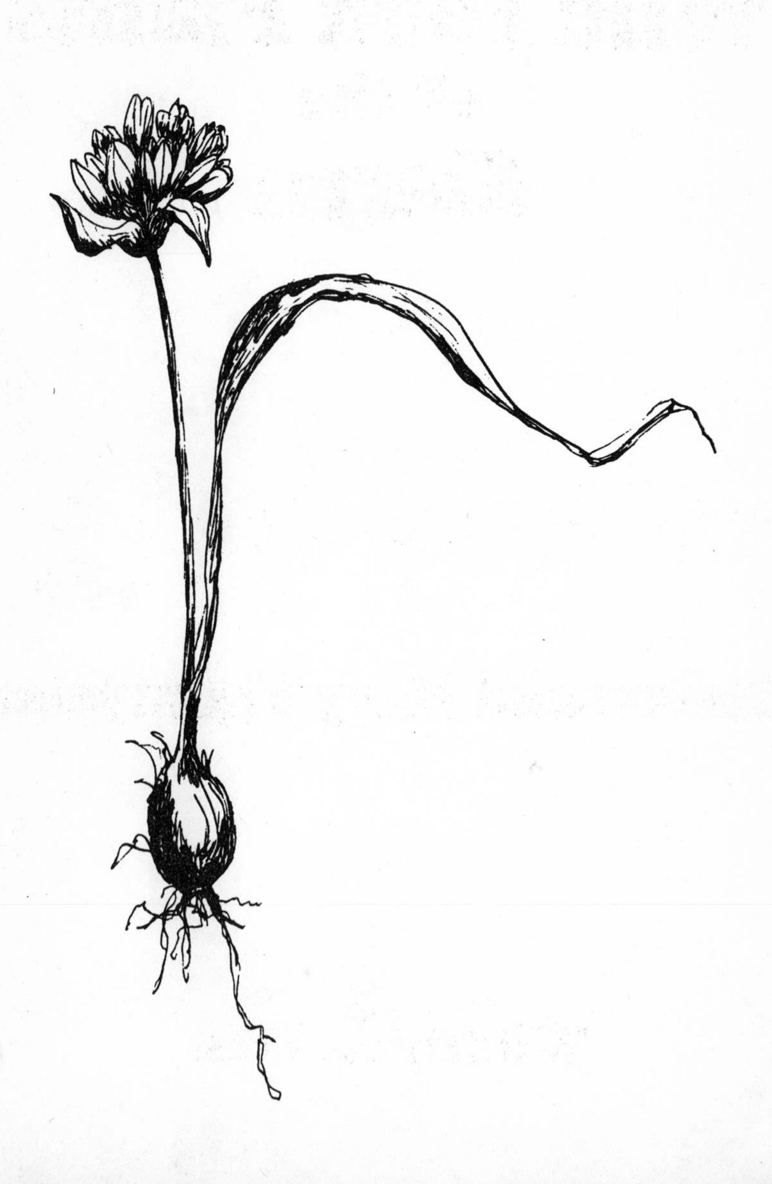

Wild Food Plants of the Sierra

Steven and Mary Thompson

Wilderness Press

ISBN 911824-52-9
Library of Congress Catalog Number 76-47776
Manufactured in the United States of America
Revised seventh printing, November 1976
Published by Wilderness Press, 2440 Bancroft Way,
Berkeley CA 94704

Cover: Big-leaf maple bark

PREFACE

This book was conceived in the mountains and is designed to be used in the mountains. The cooking instructions for wild foods are mostly appropriate for a campfire or small camp stove, with a single light-weight pot as the utensil. Mary made the drawings in the mountains from fresh plants, not from pressed, dried herbarium relics. Likewise, Steve did most of the writing during our wanderings in the Sierra.

So many people have lent assistance and inspiration in the making of this book that we cannot mention them all, but special thanks are due Stan and Barbara, Mike, Mr. and Mrs. Myers, Bill and Nancy Peppin and the Lawler family, Jeff and Lynne Dozier, and Russ Gale. Norman Clyde, the mountaineer-naturalist, gave us our first introduction to many of the plants discussed in this book.

CONTENTS

“Do you not think that all life
comes from the mountain?”

A Native American to C. G. Jung

A WILD–FOOD–PLANT PROFILE OF THE SIERRA

For someone familiar with the experience of wild places, the desire to learn the taste of wild things grows as a natural extension of this experience. The Sierra Nevada offers a wide variety of savory wild plants for the forager, the nature lover, and the wanderer.

A look at a profile across the central Sierra reveals many different plant communities, depending largely on altitude. In terms of wild food plants, the richest life zone of this mountain range is the mixed - coniferous forests and clearings of 3000 to 7000 feet on the west slope (the Sierra crest runs roughly north-south). This zone is commonly referred to as the yellow-pine belt, but it is characterized more by the great variety of trees and other plants than by any one species. The largest trees are found here – giant sequoias, sugar pines, yellow pines, incense cedars, white firs, douglas firs, black oaks. The Indians who once occupied the west slope of the Sierra traveled up and down with the seasons, but they did most of their gathering in this zone. Here they collected acorns, sugar-pine nuts, berries of all kinds, yampa tubers, harvest brodiaea bulbs, fresh greens, hazelnuts, cherries, plums, mushrooms, and many other wholesome foods.

Below the mixed-coniferous forests, the Sierra foothills are dry, especially in summer, and the vegetation is less green and lush. Still, some interesting wild food plants grow here, including digger pines, yerba santa, extensive patches of blackberries, and various bulb plants. Poison oak is a prevalent problem for the susceptible forager in the foothill woodlands.

The red-fir forests immediately above the mixed-coniferous forests are bleak places to look for wild foods. Looking like haunted places, these forests have little plant life other than the stately red firs. Douglas squirrels may love fir nuts, but they taste like turpentine to us.

Above and beside the fir forests, the lodgepole-pine belt (up to about 10,000 feet) has quite a variety of wild food plants. But the picking seems thinner at these higher elevations, and the tidbits are usually smaller. We are especially interested in food plants of the high country, though, because we like that part of the Sierra so well. Among the lodgepole pines, we find such food plants as serviceberries, onions, garlics, alpine gooseberries, golden brodiaea bulbs, wild strawberries, yampa, and alpine sheep sorrel.

Timberline is not a line but a zone, extending from the top of the forest up to the last scrub of whitebark pine. Wild plant foods do not come in quantity here, but there are a variety, including whitebark-pine nuts, mountain sorrel, alpine prickly currants, Sierra garlics, and mountain pennyroyal.

On the east side of the Sierra, the food plants found at higher elevations (above 7500 feet) are much the same as what you find on the west side except that they are sparser because the east side gets less precipitation. Below 7500 feet, the piñon-juniper woodlands compare with nothing found on the west side. Boggy aspen groves within the upper piñon belt offer a complete contrast in this dry land of piñon nuts with thimbleberries, currants, onions, and great patches of rose hips. Below the piñons is the sagebrush desert.

The elevation ranges of plants are higher in the southern Sierra and lower in the north. For example, sugar pines range from 3500 to 6500 feet in the north and 4500 to 9000 feet in the south. Anyway, plants do not really organize themselves into belts or zones so much as people organize their thinking

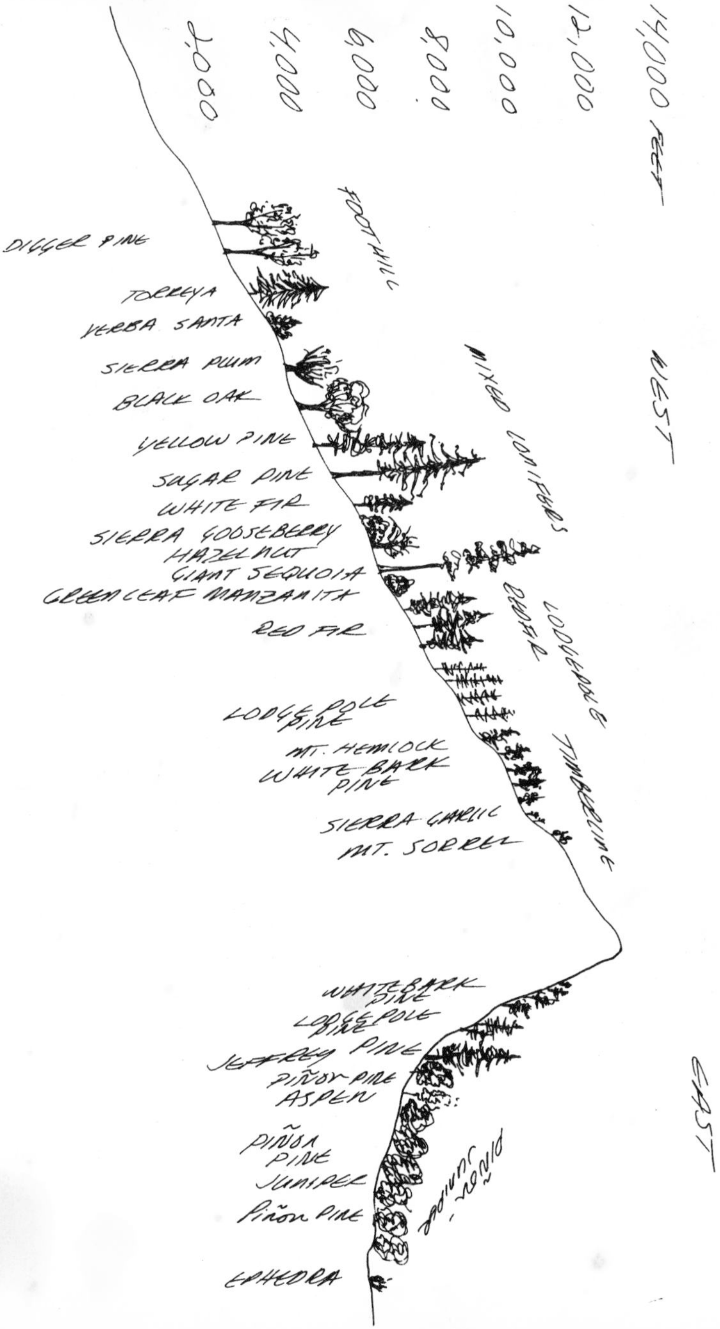

14,000 FEET
12,000
10,000
8,000
6,000
4,000
2,000

WEST

FOOTHILL

DIGGER PINE
TORREYA
YERBA SANTA
SIERRA PLUM
BLACK OAK

MIXED CONIFERS

YELLOW PINE
SUGAR PINE
WHITE FIR
SIERRA GOOSEBERRY
HAZELNUT
GIANT SEQUOIA
GREENLEAF MANZANITA

RED FIR

RED FIR

LODGEPOLE

LODGEPOLE PINE

TIMBERLINE

MT. HEMLOCK
WHITE BARK PINE
SIERRA GARLIC
MT. SORREL

EAST

WHITEBARK PINE
LODGEPOLE PINE
JEFFREY PINE
PIÑON PINE
ASPEN

PIÑON JUNIPER

PIÑON PINE
JUNIPER
PIÑON PINE
EPHEDRA

about plants this way. So don't be upset when you find lodgepole pines growing in the mixed-coniferous forest and yellow pines at timberline.

Valleys are nice places for variety in wild Sierra foods because they have a sunny (north) side and a cool (south) side, since most valleys run east-west in the Sierra. Thus, plants liking moist shade will be found on the south side – wild ginger, thimbleberries, Sierra currants, raspberries, mountain sorrel. Plants preferring dry, sunny slopes grow on the north side – harvest brodiaeas, yerba santa, Sierra garlics, soap plant, pennyroyal, squaw currants. And you can often find a ripe berry on the sunny side weeks sooner than one of the same species on the other side of the valley.

Pure mountain meadows are generally not the best places to look for wild foods, nor are dense forests. Open forests, forest edges, clearings, streamsides, rocky ledges, sandy and gravelly places, and boggy groves have more to offer. The best places of all are areas in the mixed-coniferous forest zone which have been disrupted by humans – old roadsides, logged areas, burned areas, old mines, abandoned orchards, power-line clearings, sites of old constructions – where such plants as elderberries, fireweed, yerba santa, hazelnuts, milkweed, gooseberries, and morels are quick to take advantage of the new conditions. The deer and bear also find more to eat in these disrupted areas than they do in the mature forests.

Conveniently for us foragers, man in the last one hundred years has seen to it that there is hardly a spot in the Sierra that has not in some way been disrupted. Yosemite Valley, to cite a well-known example, has been plowed, farmed, grazed, logged, burned, drained, driven upon, quarried, and constructed upon again and again, and the process is still going on. Most of the accessible mixed-coniferous forest of the Sierra has been completely or selectively logged, and modern equipment continues to make new areas accessible, hence logged. The foothills were turned upside down and sifted by the gold rushers. And almost the entire high country was overrun with sheep less than one hundred years ago. In fact, even John Muir once worked as a sheepherder for a groovy outdoor job in the mountains. Ecologists believe that the plant life has not yet recovered from the sheeps' grazing. The native bighorn sheep have hardly begun to regain their footing since their domestic relatives drove them out of house and home. Gooseberries and currants in the Sierra have long been subject to extermination efforts due to their association with white pine blister rust. If you happen to be camping in the mixed-coniferous zone in a national forest, don't be alarmed when a hèrd of cows wades across the stream to join your picnic. Meanwhile, new roads, hotels, and aerial tramways are being planned on the logic that millions of citizens would otherwise be deprived of experiencing the natural beauty that would otherwise have been there.

Well, I shouldn't sound so sarcastic because, as I said, most of these disruptions further our own special pursuit, looking for wild food plants in the Sierra.

We have often wondered about the ecological impact of seeking wild plants as food. Certainly there is no question that a sensitive person today would not destroy a mariposa lily to eat its bulb. Picking a piñon nut or an elderberry, on the other hand, does not seem to raise a measurable issue. And some of our favorite wild food plants are not a "natural" part of the landscape themselves but were introduced from Eurasia. But it is far better to be overcautious than to risk hurting a plant species in any area. We have tried to include cautions with the descriptions of wild food plants in this book. It might be said that this book is not about wild gourmet feasts but about tastes and nibbles.

When the Sierra was inhabited by Indians, they were an important influence in the ecological balance. They kept the deer in check, they gathered great quantities of wild nuts and bulbs and other food plants, and they deliberately set fires to clear out the underbrush and improve conditions for certain food plants. But the Indians were aware of their place in the environment and lived in the Sierra for thousands of years without destroying the beauty. Most significantly, they were aware of the dangers of overpopulation and practiced birth control. Their medical lore included a number of oral-contraceptive plants. At least one of these species *(Lithospermum ruderale)* has been tested in a modern laboratory and found to suppress ovulation in rats. Regarding food plants, the Indians had a custom of not taking from the first plant of a species one came upon. Rather, one should ask its help in finding more of the same. This custom turns out to be sound conservation practice. And not only does it prevent you from wiping out a solitary specimen, it often leads you to many plants that you otherwise would carelessly have missed.

With the Endangered Species Act of 1973, Congress gave recognition to the fact that some plants are in serious trouble with our society. Preliminary lists of such species have been published in the Federal Register; however, work is still in progress and these lists must be regarded as incomplete. Regional Forest Service offices should have the most current information on species being considered for Threatened or Endangered status in their area.

Identification is usually the biggest obstacle to someone interested in learning about wild food plants. There are no general rules for distinguishing an edible plant from an unsavory or poisonous one. One simply must identify a plant correctly before trying it. Once you have tasted an edible plant, however, you are not likely to ever forget it. We include warnings throughout the book on poisonous plants which might possibly be confused with edible ones.

The easiest way in which to learn about the plants of the Sierra is to travel with someone who can identify the plants and point them out to you. You can check his knowledge against a book later. Even a botanist could not possibly know every species of plant in the Sierra. Botanical gardens with native species are also good places to learn the names of plants.

Many food plants you can identify with this book alone, but we urge you to become familiar with other plant books as well. There are a number of books on plants of the Sierra that we have found useful. *Sierra Nevada Natural History* by Tracy I. Storer and Robert L. Usinger is a convenient paperback with a fair number of plants covered. *California Mountain Wildflowers* by Phillip A. Munz, also in paper cover, does not limit its discussion to the Sierra but is nonetheless useful to have here. Neither of these well-illustrated books is anywhere near complete, however. The definitive work covering plants of the Sierra is *A California Flora* by Phillip A. Munz and David D. Keck. This work is a technical flora with very few illustrations, but we encourage you not to be overawed by it, even if you haven't had a course in taxonomy, as the authors provide a glossary and other explanatory matter. The completeness of this tome avoids some of the frustrations encountered with the handier, but incomplete, illustrated guides. An illustrated glossary, such as found in *How to Identify Plants* by H. D. Harrington and L. W. Durrell, makes a good companion to *A California Flora.*

A plant guide is really just a maze through which one has to travel to learn a plant's name. Once you know it, a plant is usually as easy to recognize as a person.

SUGAR PINE

Sugar pine *(Pinus Lambertiana)* nuts were considered a real delicacy by the Miwok Indians of the west slope of the Sierra. These nuts are indeed good. They are also one of the most challenging of wild foods to collect.

A typical autumn scene finds me out on a limb a hundred feet above the ground bouncing vigorously. The long cone at the very end of the limb flops up and down until I give the bounce a circular motion. The cone whirls around and around, twisting itself loose. It sails away through the air and lands, of course, in a dense thicket of whitethorn ceanothus. Mary, watching from the ground, tries to keep track of where each cone lands – not always possible while also keeping out of the way.

Climbing down out of the tree seems more difficult than climbing up as the branches become farther apart and the trunk becomes thicker. Eventually I have only dead limbs to grab and then none. The trickiest part is transfering my weight to the dead lodgepole pine pole I had leaned against the trunk. Then it is easy to shinny on down to the ground.

The largest of pines, a mature sugar pine is an inspiring sight. In the Sierra they live between 3500 and 9000 feet in the mixed-coniferous forests, scattered among other pines, firs, incense cedars, and sequoias. The Sierra has the largest sugar pines anywhere, eight feet thick and two hundred feet in height. Often there are no branches, even dead ones, for a considerable distance up the trunk. The smaller,

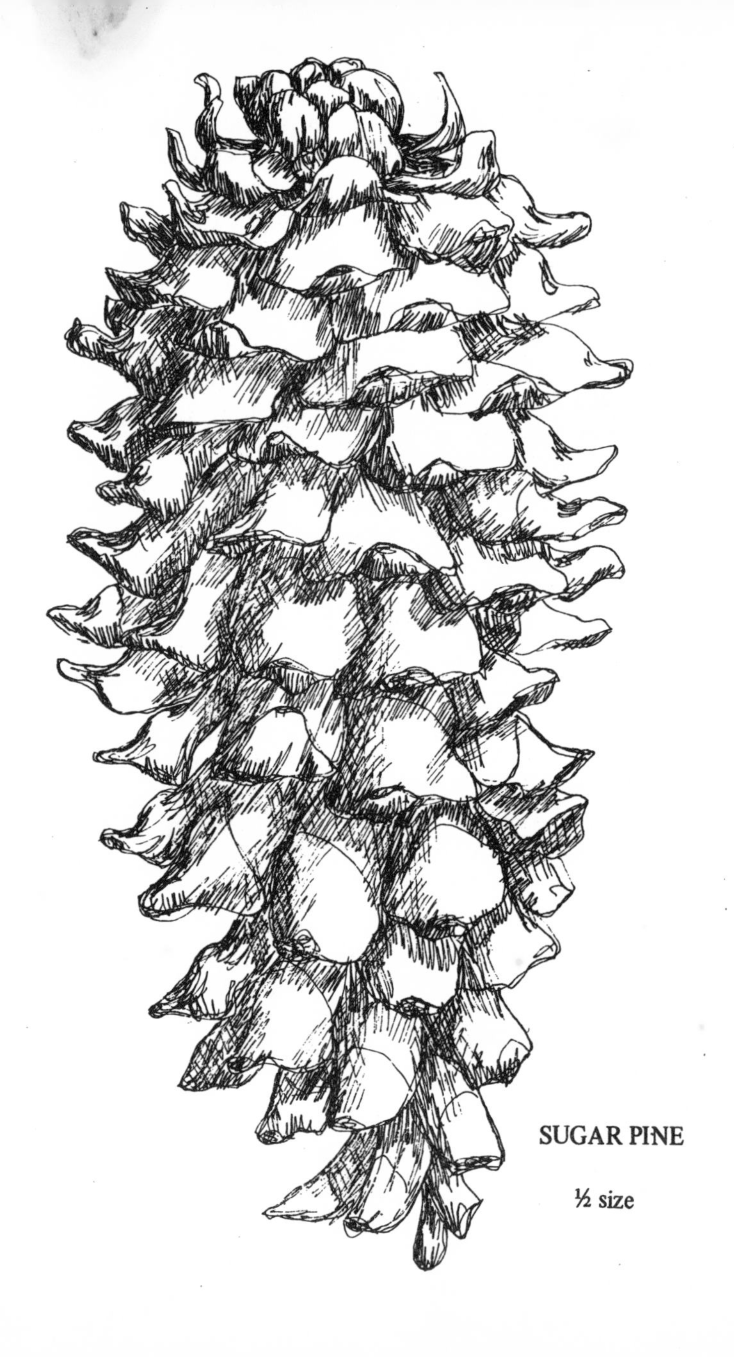

SUGAR PINE

½ size

easier-to-climb sugar pines are much less likely to have any cones. Occasionally, you can find a smaller tree, usually a white fir, growing next to a sugar pine which gives you access to the lower branches. Or, if you can find a twenty or thirty foot pole nearby, you can lean it against the base of the sugar pine as I did. The Miwoks often used this shinny-pole method.

Sugar pines can easily be recognized by their long cones. These cones are about a foot and a half long and hang from the ends of the branches. You cannot climb out to them, but must use your hands or feet on the branch to twist them off by bouncing. The nuts are ready to collect when the cones are a mature brown but haven't yet opened. The right time usually arrives in mid-September. If you wait too long, the scales of the cones open, and the winged nuts fly away in all directions like little helicopters.

One September, we were looking for sugar-pine cones in a particular forest where we couldn't find a single climbable sugar pine. In the distance we heard some loud crashes and thuds and went to investigate. There we found a Douglas squirrel cutting cones from a towering sugar pine. We couldn't restrain ourselves from exploiting his labors by stealing a couple of cones, but in the long run the saving of effort was not worth the pangs of guilt. For someone with no qualms about robbing a squirrel, however, this would certainly be the ultimate method of collecting sugar-pine cones.

The cones are extremely pitchy. Place them in a pile of pine needles in a fireplace and set the needles afire. Turn the cones as necessary until the pitch is seared off and the scales are loosened.

Sugar-pine cones are large, and the scales tightly interlocking. It is very hard to peel them back for the nuts. Again the Miwok system is most practical. Take a stone or hammer and hit the tip of the cone (the end away from the part that attaches to the tree) until it splits. Then, with your hands, finish splitting the cone into halves and fourths lengthwise. For some reason, the cone does not agree at all to being split from the other end.

The scales are then easy to peel back, and the nuts come right out. These sweet-tasting nuts are ready to be shelled and eaten. Some trees produce bigger nuts than others, but they are all smaller than piñon nuts. We like the flavor of sugar-pine nuts better.

Each nut has a long wing attached to it. Many of these wings end up mixed with the nuts. If you want to separate the nuts from the debris, the Indian method of winnowing works well. Using your hands or a shallow open container, toss the nuts repeatedly into the air and catch them again. Any light breeze will take the wings away.

Parching the nuts may be necessary in order to prevent mold. Parching can be done by a light scorching in a hot frying pan. This process is described in more detail in the next chapter.

Sugar-pine needles grow in bundles of five. A small handful of these needles, steeped in boiling water, makes our favorite pine-tree tea.

John Muir described the sap of the sugar pine, which accumulates around scars in the bark, as his favorite sweetening. We have tried it on a number of occasions and found that it tasted very much like pitch or turpentine. Perhaps we haven't found the right season or the proper kind of scar. In any case, the sap is also reported to be a laxative.

Sugar-pine cones are well known as decorative objects. They can be seen for sale in resort gift shops, painted with shiny varnish, of course. Many times we have seen tourists packing sugar-pine cones into their cars, presumably to decorate their mantelpieces.

Unfortunately, sugar pines are one of the most valuable timber trees of the Sierra, and virgin stands are still being attacked by the lumber industry. A few years ago we hiked past a place called Sugar Pine Hill. Later, I heard that the timbermen were pushing hard to harvest this virgin stand. I wonder if they plan to change the name of the hill, too.

PIÑON PINE

Every species of pine has edible nuts. The main factor which makes some species preferable is the size of the nuts. For illustration, try collecting lodgepole-pine nuts. If you are even able to shell one, your dexterity compares well with a chipmunk's. The one-leaved piñon *(Pinus monophylla)*, on the other hand, has the largest nuts of any pine. They are among the most profuse and easiest to collect, and they are easy to shell and taste good. They were once the staple food of the Paiutes who lived on the east side of the Sierra. Piñon nuts are one of the few wild foods which we collect in quantity.

To us, September always means heading for the piñon woodlands, which grow up to about 8000 feet on the slopes and ridges east of the Sierra crest. These woodlands, which form the transition between the mountains and the desert, are interesting places to spend a few days. Juniper and desert mahogany grow among the piñons, and aspens, water birch, and red dogwood line the streams. There are usually other wild foods to collect, such as rose hips and ephedra. We have also seen a very few piñon trees on the west side of the Sierra at about 5000 feet, but we have never collected the nuts there.

Just as sugar pines have five needles in a bundle, one-leaved piñons have one needle in a "bundle". These needles are round in cross section and tapered to a strong, sharp point, as you will notice if you climb the trees in a light shirt. The foliage is a pale smokey-green. The trees are small relative to other pines and are roundish in form. On the eastern side of the Sierra, these pines can be seen covering whole hillsides in overwhelming numbers for miles in all directions.

When the nuts ripen within the cones, the scales of the cones open and let the seeds drop, one by one. It is possible to collect the nuts from the ground at this time. You will get more if you shake the tree branches. However, we have not found this method very productive as the seeds are scattered in the underbrush, and many have already been claimed by worms and squirrels.

We find it far more efficient to gather mature cones while the nuts are still inside. When the green cones become brown at the tips of the scales, the nuts within should be mature. This stage occurs from early September to early October, depending on the area. Also, we have seen the season arrive at a particular place two weeks later than it had the year before, due, perhaps, to weather conditions.

Piñons are notorious for giving a bumper crop one year and nothing the next. One year, for example, we returned to an area which had been especially prolific the year before. Instead of thousands of cones, we found three in the whole area. Two canyons to the north, however, we found piñons with a moderate number of cones. Interestingly, the piñons everywhere seemed to be having an off year, while sugar pines had the most cones we'd ever seen.

Before selecting a tree to collect cones from, it is worth the effort to pry back one scale to inspect the quality of the nuts. Nuts vary considerably from tree to tree, but are quite consistent on a single tree. We look for trees with extra large nuts.

If a nut has a worm in it, chances are that other nuts of the same cone and other cones of the same tree have worms. These insect larvae are not appetizing to look at, and it is even worse to think of burning them alive. They always leave plenty of good cones for us, so we see no reason to persecute them.

PIÑON PINE

life size

Therefore, it is certainly worth looking at each cone as you pick and leaving the inhabited ones behind. The worm holes can almost always be spotted by the sawdust around them.

We gather the cones by climbing the trees and throwing the cones to a clear spot of ground. Sometimes, we throw them onto an old tarp which we then use to carry them. Piñon trees present no problem to climb except that they are prickly, dense, and pitchy. Over the years, we have learned that it is more pleasant to wear a hat than to wash pitch out of our hair with peanut butter and gasoline. The Indians sometimes used a pole with a hook on the end of it to pull the cones down without having to climb the tree.

The next step is to gather the cones and take them to a safe spot to built a fire in this dry country. You want to burn the pitch off the outside of the cones to unglue the scales so you can get at the nuts. Usually, some Jeffrey pines can be found growing in the area. Jeffrey needles (three in a bundle) are long and make a better fire than piñon needles. Gather a good amount of dry needles, mix the cones with them in the fireplace, and light the fire. Poke the cones around to see that they get scorched on all sides. This hot, fast fire should be just right to burn away the pitch without burning the nuts inside. It is also possible to use small twigs instead of needles or to build a wood fire and place the cones on a grate.

The pine needle fires can make a conspicuous volume of smoke, and you should check the fire permit regulations of the area. One year, we were processing piñon cones in a Forest Service campground. After several hours of billowing clouds of smoke, a Forest Service man came driving up with a look of great urgency. "Hey, is that just a campfire?" Apparently the smoke had been spotted from a forest fire lookout. When I answered yes, it was just a campfire, the look of urgency turned to one of serious concern. "Uh – having trouble getting it started?"

With the cones out of the fire, simply bend back each scale, popping out the nuts. There are normally two nuts under each scale. The number of nuts per cone varies from about ten to sixty. The nuts are surprisingly large for the size of the cones. They vary in color from light tan to dark mahogany. If you have many cones to do, it is a good idea to wrap adhesive tape around your thumbs to keep them from getting sore.

If you have a van or pick-up, it is possible for you to skip the burning and just take the cones somewhere to dry. The cones tend to open by themselves and the nuts can be pulled out. We know people who prefer the flavor of these raw nuts. Personally, we like the slightly roasted flavor the nuts get when the cones are scorched. In either case, we like the nuts best after they have had some time to dry.

If the weather becomes at all damp, fresh pine nuts are extremely susceptible to mold. Spread them out as often as possible until they are dry, and store them in cloth or paper bags or boxes rather than in plastic or glass. It may be desirable to parch the nuts to prevent mold or to destroy mold if it appears. The Indians mixed hot coals with the nuts in a basket and shook it around to parch the nuts. We have parched them by placing them in a hot frying pan until the shells start to pop like popcorn. If mold appears and you kill it soon enough, all is well. But if mold spreads to the nutmeats inside, the flavor is ruined.

Indians made flour, soup, baby food and nut butter from pine nuts. They shelled the nuts in mass between two flat stones. We have been fairly successful with this method, using a board and a rolling pin. Our favorite way of eating pine nuts is like peanuts, so to speak, shelling and eating them one at a time. We also particularly like them cooked in whole-grain brown-rice dishes. When put through a grinder, pine-nut flour is a delicious addition to homemade bread. I am especially fond of pine nuts in stuffed green peppers.

Piñon needles, like the sugar-pine needles, make a surprisingly nice tea. In fact, every kind of pine needle – and the needles of some other conifers, such as mountain hemlock *(Tsuga Mertensiana)* and Douglas fir *(Pseudotsuga Menziesii)* – can be used to make a refreshing tea. Each species has a flavor distinctly its own. Our favorites are sugar pine and Jeffrey pine. Contrary to what you might expect, the tea does not taste like turpentine, but is rather tangy and lemony. However, you might first want to check that one of our government agencies hasn't been spraying the trees with insecticide.

The inner bark of pines was used as an emergency food by Indians, particularly in lean winters when little other food was available. The inner bark fibers were beaten into a flour and cooked into cakes. We have not tried it, and we have not heard that it is particularly savory.

We have heard alarming stories of piñons being cut for firewood and to increase grazing for livestock. It is certainly upsetting to think of such a charming tree being laid to waste because of a change in economies.

WHITEBARK PINE

As you climb above timberline in the Sierra, the last trees you see will probably be whitebark pines *(Pinus albicaulis).* At the upper limits of the timberline zone, these low, twisted, windswept forms with weathered dead snags sticking out give the Sierra high country a unique beauty. Lower down, where the scattered trees and shrubs of timberline give way to full forests, whitebark pines grow into tall, sturdy trees, often with forked trunks. Their range is between 8000 and 12,000 feet in the Sierra. The needles grow in bundles of five.

One of the tastiest varieties of pine nuts grows on the whitebark pine. These nuts are the favorite food of the Clark's nutcracker. As early as July, these alpine crows can be seen noisily perched in the treetops pecking at the unripe cones for their half-developed seeds. Chickarees and ground squirrels of the timberline zone are also fond of these nuts.

Whitebark-pine nuts are an especially welcome wild treat in the high subalpine country where food gathering is relatively lean. The nuts are smaller than piñon nuts, but large for the size of their cones. The shells are a beautiful purple color.

The nuts are generally ripe by late August. When they fall in September, the scales fall with them and the cone disintegrates on the branch. The cones tend to grow at the very tops of the trees. To get at them you can either climb a tree or hike to a higher elevation where the "trees" are only six feet tall and you can reach the cones from the ground. Larger whitebark

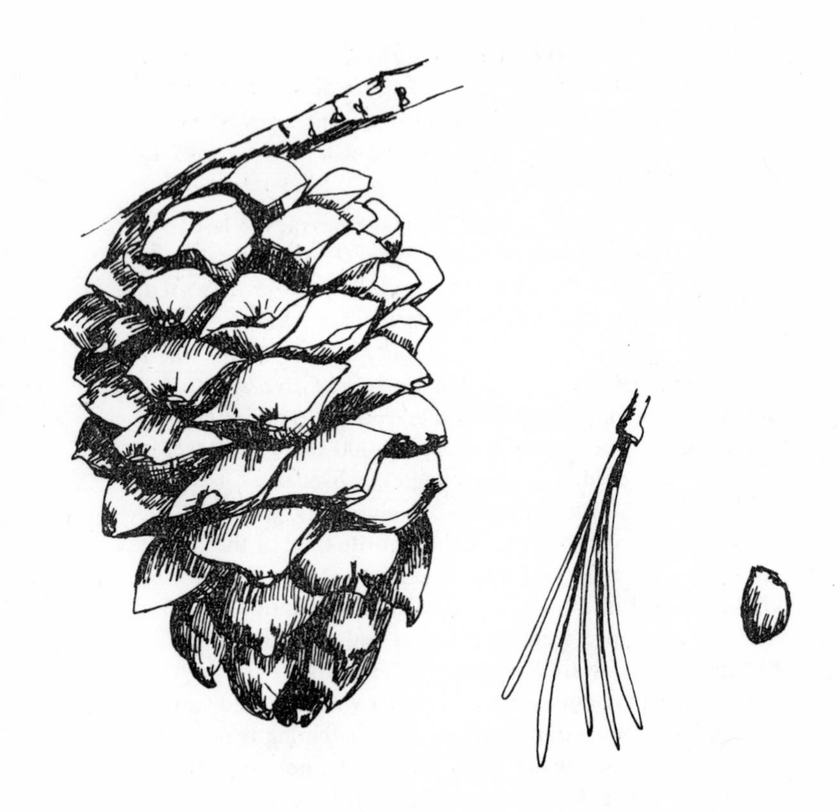

WHITEBARK PINE

life size

pines often have forked trunks and are usually easy to climb using cross-pressure between the trunks.

As with piñons, watch carefully for worm holes in the cones.

The pitchy fresh cones are processed in much the same manner as piñons except that the long needles of Jeffrey pine will not be available. Whitebark or lodgepole needles make a poor substitute. So either gather many small twigs for burning the pitch off or put the cones on a grate over the campfire.

The needles can be brewed into a good tea.

Because of their small size and their importance to the native animals at this altitude, we wouldn't try to gather a large quantity of these fancy nuts. But for a delightful snack eaten right where you collect it, a gourmet catering service couldn't treat you better – especially at 11,000 feet.

DIGGER PINE

Nuts of the digger pine *(Pinus Sabiniana)* were one of the most important foods gathered by Indians in the Sierra foothills and the Coast Ranges. The nuts are large – longer but narrower than piñon nuts. The flavor is good and very different from other pine nuts we have tried. The shell is very hard and almost impossible to crack between the teeth.

Digger pine cones are large and heavy with sharp points on the outside of the scales. The trees can be fairly hard to climb and the cones a struggle to twist loose. Old cones which may have dropped their seeds some years before often still cling to the trees.

The nuts are extremely difficult to extract from the fresh cones. The scales are thick and stiff and do not come unglued from each other when the outside of the cone is scorched in a fire. We processed a few of these cones using a hammer, chisel, and pliers with much effort.

A more feasible method is to let the cones dry until the scales separate by themselves. Then shake and beat the nuts out, or pry them out with a knife if necessary.

The needles are long and grow three in a bundle. These irregularly-shaped trees with forking trunks and grey-green foliage are perhaps the most integral part of the Sierra foothill scene.

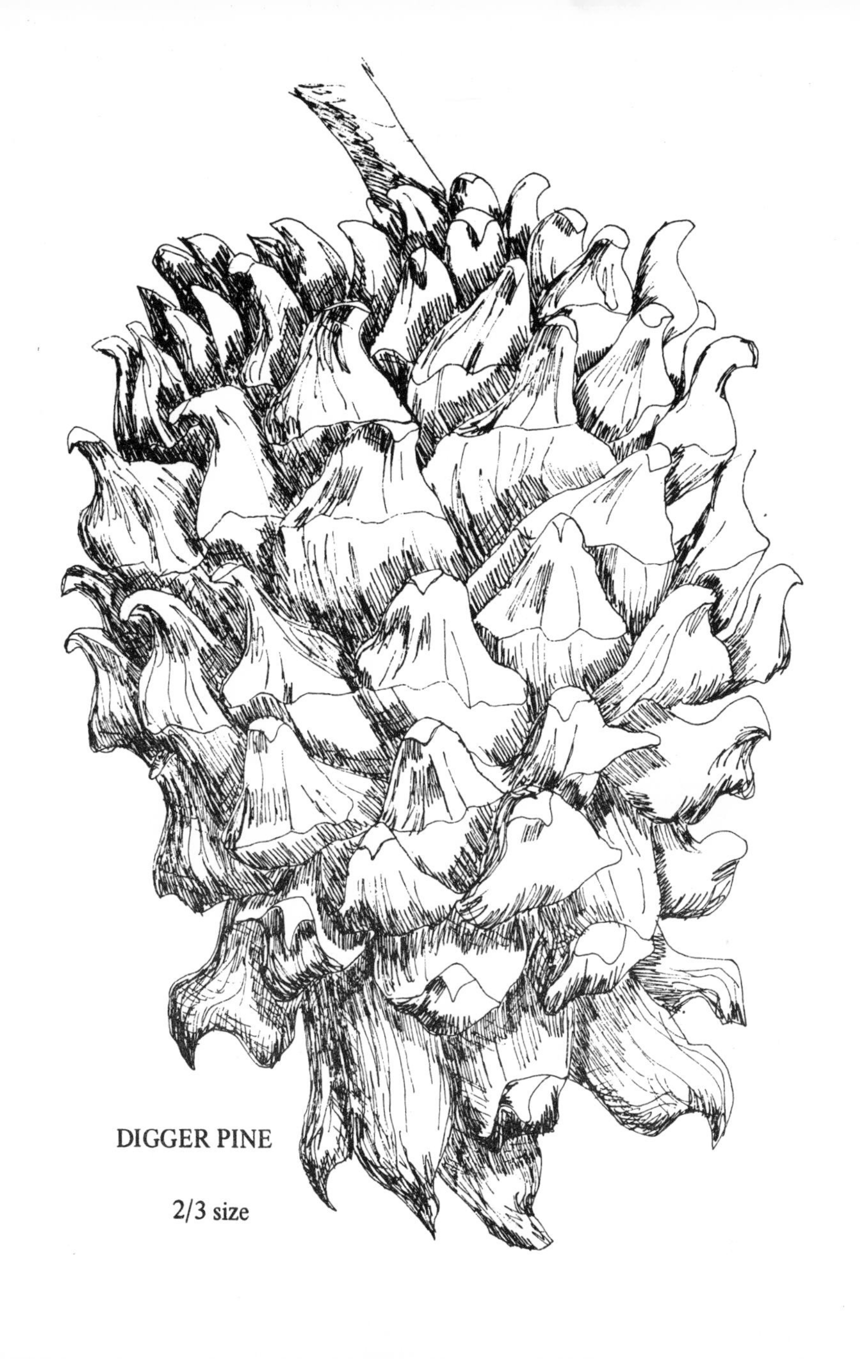

DIGGER PINE

2/3 size

TORREYA

An unusual conifer of the Sierra is the torreya *(Torreya californica),* also known as California nutmeg. It is a fairly small tree which rarely grows taller than fifty feet. It grows in cool, shaded canyons of the west slope of the Sierra below 4500 feet.

A noticeable feature of the torreya is the stiffness of the needles, which turn into real needle-points at the ends. You may one day find yourself being stabbed from all sides as you climb out of a canyon. This evergreen is definitely not recommended for bough beds, even aside from conservation considerations. The needles grow in flat sprays. They are dark green above, yellow-green below with two light grooves.

The fruit is green with wavy lines on it and contains a large, ellipsoid nut with a hard shell. If you cut the nutmeat in half you will see that the central part is white and the outer part is reddish. The two parts interlock in a convoluted pattern. You can eat the white part. It tastes like coconut and is very good. But the reddish layer is very astringent, drying up the saliva in your mouth. Unfortunately, the convoluted structure makes it difficult to separate the two layers.

Indians ate the whole nuts after thorough roasting and reportedly were quite fond of them. We tried roasting the nuts in an oven at 350 degrees for fifty minutes. Much of the astringency was thus removed, and the roasted nuts tasted

TORREYA

life size

good at first. However, after a couple of minutes a delayed astringency reaction still took place, leaving us rinsing our mouths out with water – to little avail.

BLACK OAK

Acorns were the staff of life for the Monaches and Miwoks living on the west side of the Sierra. Their favorite source was the California black oak *(Quercus Kelloggii).* Black oaks grow among the conifers up to about 7000 feet on the west side of the Sierra. These noble trees grow to great size in the valleys and forests of 4000 to 5000 feet.

Acorns from other kinds of oaks were also used for food. But in general, other varieties are more bitter and require more processing. When the acorn crop failed completely, the Indians could fall back on the large, round nuts of the California buckeye *(Aesculus californica).* They had even more bitterness to be leached out, however, and were never a preferred food.

The black-oak acorns usually ripen in early September. In some years, there is a great abundance of acorns, while in other years, the trees produce only few. The best time to collect acorns is just when they are ready to fall from the trees and can be shaken from the branches. After they have fallen by themselves it is harder to find them on the ground, since the squirrels, blue jays and worms have been hard at work.

Acorns contain bitter tannin, so it is necessary to leach out the tannin before eating them. The Indians had a refined system for grinding and leaching acorns. The process involved many techniques, and only a sketchy outline is given here. Shelled acorns were crushed and ground with rock pestles in mortar holes. Soap-root brushes were used to sweep the meal around. The meal was placed in a scooped-out hollow in a

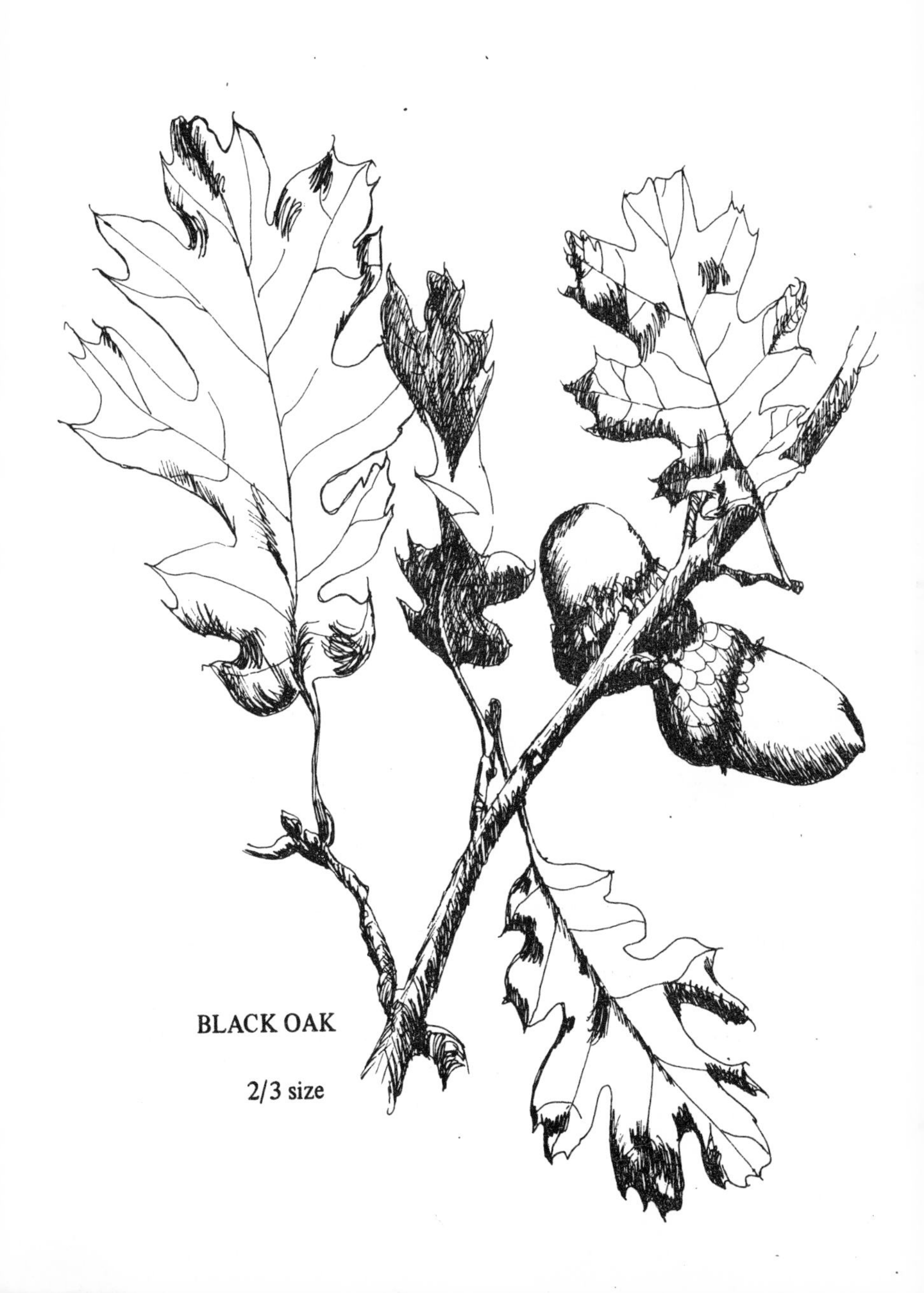

BLACK OAK

2/3 size

sandy spot. Water was poured repeatedly over the meal, and the tannin was washed out into the sand.

The mortar holes where the Indians did their grinding can still be seen throughout the Sierra. They are smooth, round holes up to about ten inches deep in flat granite slabs and on top of boulders. Polished, rounded rocks from riverbeds served as pestles. You should leave these pestles in place when you find them. We are always amazed at how common these grinding sites are in the Sierra. Most of them are found in the black-oak belt on the west side and in the piñon belt on the east. However, we have also found grinding sites well above the oaks or piñons – some above 9000 feet. They may have used these to grind various types of seeds collected locally, but it also seems likely that they carried whole acorns with them when they traveled into the high country and ground them up as needed. Whole acorns keep fresher than meal.

We have often wondered about the "backpacking" techniques used by Indians in the Sierra. Judging by the mortar holes and obsidian chippings they certainly went everywhere, including places a modern climbers' guidebook would rate "class 3" – the most difficult climbing that does not require a rope or other specialized equipment. Yet these Indians had their permanent homes at much lower elevations, usually below 4000 feet. One thing that strikes us is the superb variety of ideal backpacking foods these Indians had. In addition to the fresh food plants and game they could obtain along the way, they had all sorts of nutritious dried nuts and seeds, dehydrated vegetables and fruits of many kinds, and a large selection of vitamin-rich teas. Of course, there was also always jerked venison and dried fish.

We make acorn meal chunky style, chopping the nutmeats up with a knife or grinding them in a hand mill. Water must be poured over the chunks many times, starting with cold water and increasing its temperature to help dissolve the tannin. The coarse meal may be held in a cloth or strainer. Boiling the meal or just leaving it to soak can be a help. Taste a small amount to determine when all the bitterness is gone. The finished product can be eaten plain or used to enhance other foods such as rice dishes and bread recipes.

When we first tried acorn meal, we expected it to be a bland mush, as described in the accounts of various ethnocentric early explorers. But we found it to be very tasty and crunchy. Acorn grits are not comparable with cereal in flavor or texture, but with other nuts. Let's face it, we're the ones who have been living on a diet of bland foods.

CALIFORNIA BAY

The California laurel or bay tree *(Umbellularia californica)* can easily be recognized by its aromatic odor. This spicy aroma can be quite pleasing when mild, but if you tear a leaf and hold it under your nose, it is strongly pungent, and your sinuses will fight back.

Bay trees grow to about sixty feet in height and have a broad crown. The trunk reaches one or two feet in diameter. They grow along canyon sides up to about 6000 feet on the west side of the Sierra.

The leaves of this tree make a very good substitute for the commercial bay leaves. Use a few dried leaves in stews, soups, and meat dishes for flavoring. They can easily be dried and saved for later use.

California Indians apparently liked to roast the bay nuts and eat them. We have tried them also but have not been able to tolerate the taste. The nuts are almost an inch in diameter. They are green becoming purplish when fully ripe. The inner kernel is the main part that was used for food. We dried these and then thoroughly roasted them in an oven. To our taste, they were much too bitter and odd tasting. Very likely this was an acquired taste for the Indians, just as coffee might taste odd and bitter to someone on his first try.

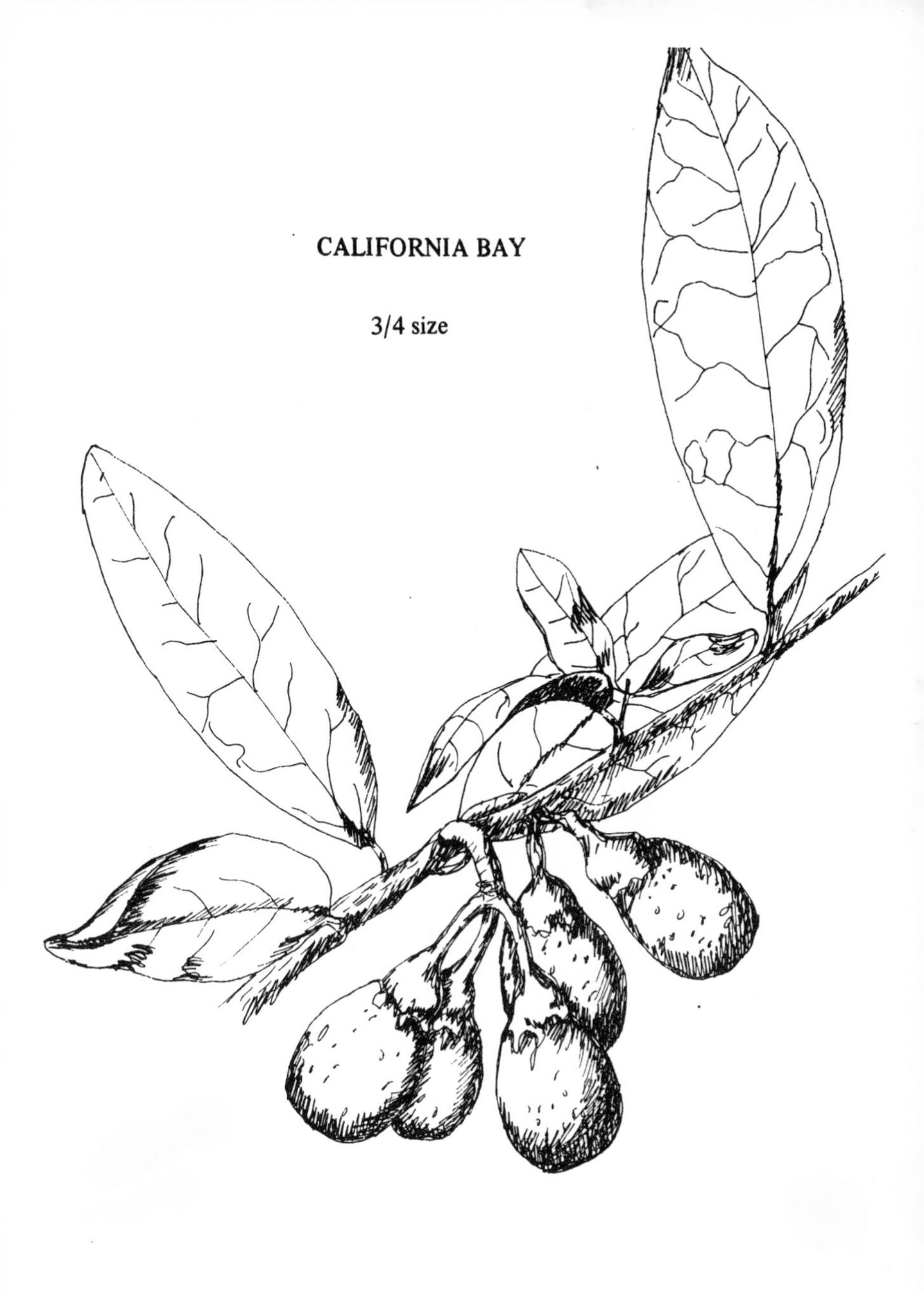
CALIFORNIA BAY
3/4 size

WILD APPLES

Wild apples? In the Sierra? Well, our discussion of wild foods of the Sierra wouldn't be realistic if we overlooked the many varieties of fruit trees that have escaped domesticity over the years and now grow wild in the mountains. Some of these trees are found in old, abandoned, mountain orchards. Others can be found growing where it is highly unlikely that any human deliberately planted them. Thus, they must be considered fully naturalized wild plants of this region.

During our wanderings through the meadows and forests of the west side of the Sierra, we have seen apple, peach, pear, cherry, plum, and fig trees of foreign ancestry growing in wild places. Most of these fruit trees are found between 3000 and 5000 feet near areas with some history to them. For example we found the fig trees growing in the vicinity of a long-abandoned mining tramway. With no help from man, fruit trees seem to thrive in the invigorating mountain climate. Much of the fruit they bear is of very high quality. We have had some busy canning seasons in September and October while camping in the Sierra.

The most common of these drop-outs from civilization are apple trees. Some sturdy and prolific specimens grow in abandoned orchards where they were planted over a hundred years ago by early settlers. Other apple trees are sometimes seen in the woods and along streams in seemingly random places. The apples from some of these trees are the best apples you will ever find nowadays, especially for cooking. By comparison, supermarket apples seem like tasteless balls of pith inflated with air. Some of our century-old, unpruned, Sierra apple trees produce apples that are huge. They are so dense and juicy that they almost split by themselves.

Black bears love the apples too and often climb out on amazingly thin limbs after them. You can see their claw marks on the trunks.

Many of the apples have worms, but the wormholes are usually confined to the cores. We consider worms preferable to poison spray, anyway. Isolated apple trees often have no worms at all, while in our experience orchard trees always have them.

Apples are very high in pectin and are ideal to add to low-pectin fruits such as chokecherries and rose hips in making jam. Conveniently, apples come into season at the same time as these other fruits in the Sierra.

Dozens of species of introduced plants are found thriving in the Sierra today, even in the most remote areas. We consider fruit trees the most welcome of non-native plants.

BIG-LEAF MAPLE

There are not many wild foods that can be gathered in the middle of winter in the mountains, but maple syrup is one. The big-leaf maple *(Acer macrophyllum)* is one of the finest trees of the Sierra. The great leaves and the winged, hairy seeds make it an easy tree to identify – except in winter when the leaves and seeds have fallen. The surest way of finding a grove of big-leaf maples is to locate them in the summer or in autumn, when the leaves have turned soft yellow. Notice carefully the handsome bark with its narrow, vertical ridges and brown to grey color. (The cover of this book is a drawing of big-leaf maple bark.) The upper branches sprout opposite one another and curve upwards into straight shoots. An occasional tree will still hold a shriveled leaf or a few of the seeds to ascertain your identification. Once you are familiar with the appearance of big-leaf maples, they are easy to recognize even without their leaves.

Big-leaf maples grow in canyons, along streams, and at the edges of coniferous forests up to about 6000 feet on the west slope of the Sierra. They are usually moderate trees in size, but sometimes noble specimens are found with trunks several feet thick.

We have had our best luck in collecting big-leaf maple syrup in the Sierra in February at 5000 feet. The weather conditions to look for are warm, sunny days following freezing nights after some real winter weather has hit. This is well before the leaves have appeared, although the new buds are much in

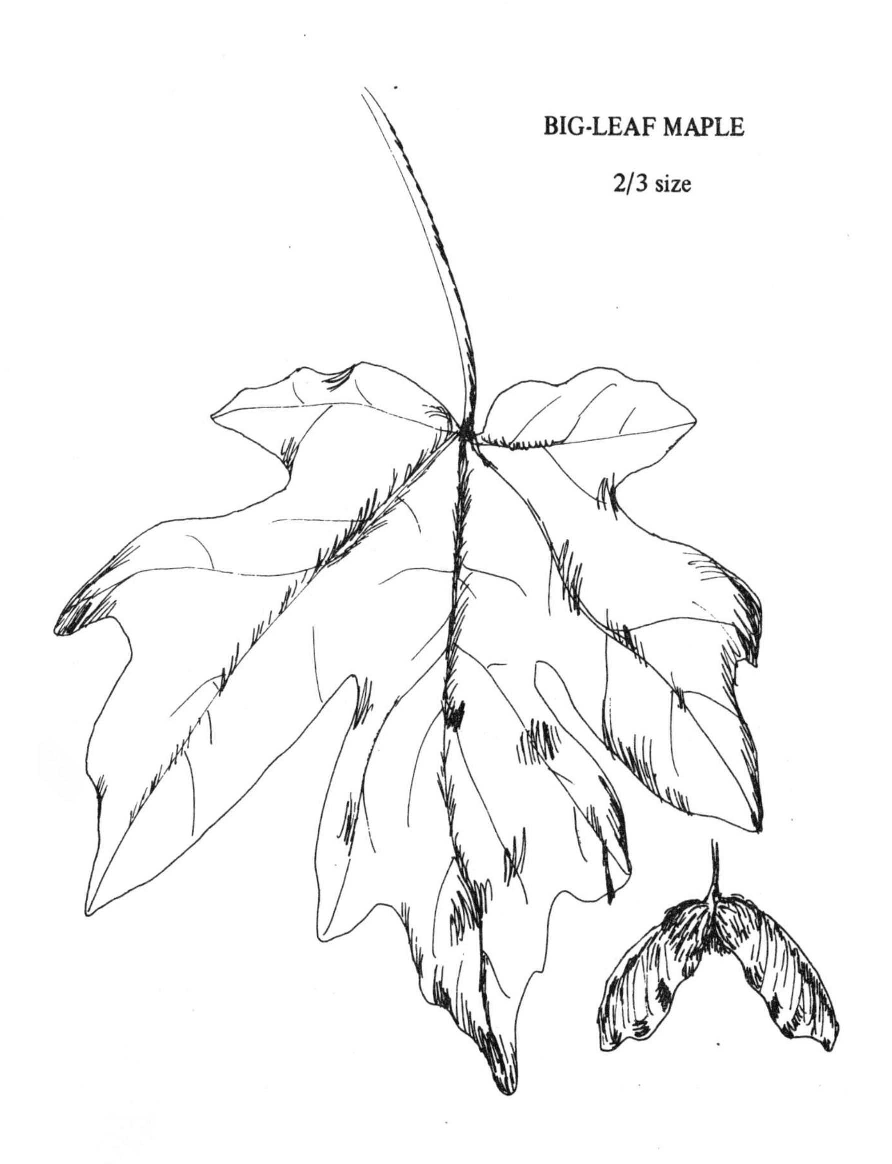
BIG-LEAF MAPLE
2/3 size

evidence. There likely could be some good maple-sugaring days in January of some years, but we have never hit upon the right conditions in the right place during that month. The trees have repeatedly refused to yield us sap in April and late March, even when the weather conditions seemed perfect and the leaf-buds had not yet opened.

When you have found a good tree on the right winter day, tap a hole into the trunk at a convenient height about two inches deep, using a brace and a 3/8 to 1/2 inch bit. The hole should slope so the sap will run out. We have always done better with a hole on the sunny side of the trunk than on the north side, although it is possible that the north side could yield more later in the season. For a spout, you can use a piece of elderberry stem which is split and the pith removed. A stick of maple or other hardwood can also be made into a spout by splitting a short length and carving a groove along the flat surface. The spout should be a tight fit in the hole. Tap it into the hole, using a mallet or the heel of your boot while you stand in the snow on one foot. The spout should be notched to hold a string or wire which suspends a tin can to catch the drippings. Once we tried saving the effort of tying the cans to the spouts and just set them on the snow where the sap would drip into them. Unfortunately, the snow turned out to harbor billions of tiny, black, hopping insects, and we felt compelled to discard the bug-speckled sap thus collected.

If a tree is ready to flow sap, it will let you know almost immediately after you have drilled. Within a minute the inside of the hole should be moist and the sap will soon be oozing out. Even when all factors seem equal, we have found some trees give much more sap than others, even trees right next to each other. A good tree can be tapped in more than one spot at once, but you should avoid overtaxing it with more than two or three holes.

We always seal the holes with wooden plugs carved from sticks when we are through to protect the tree from infection and from further loss of sap while it heals. From the experience of generations of sugar-maple tapping in the northeast, it does not appear that tapping a tree is detrimental to its long-range health. However, it is a good idea not to tap one tree heavily in successive years or to tap it in too many places at once.

We have never managed to collect the gallons and gallons of sap people tell of collecting from other species of maples back east. The sap goes drip, drip, drip, and a small-size can has sufficed on our taps.

However, the maple syrup we have made from the sap is the best maple syrup we have ever tasted. The sap as it comes from the big-leaf maple is the color of apple cider. Put it through a fine strainer to remove the pieces of bark, odd insects, or whatever else might have fallen into the can. The sweetness can hardly be tasted in the natural sap. The sap must be boiled down to a small fraction of its original volume until the taste is of the strength desired for maple syrup. Boil it slowly over a low heat, being careful not to scorch it when it is nearly done. Taste a small amount often until you are satisfied (an understatement!) with the flavor. We have boiled the sap until is was about one 20th of its original volume, and the resulting syrup was a rich, dark brown color and more flavorful than commercial maple syrup.

SIERRA PLUM

Wild plums should be one of the best wild foods to collect in the Sierra, but in the areas we have observed, most of the fruits are destroyed by a fungus. A tree forms thousands of beautiful blossoms, and a month or two later it is laden with shriveled, hollow things which should have been plums. These crumbly, bladdery structures do not even contain a pit. The few healthy plums we find are fine fruits about one inch long and three-fourths inch across. They ripen red-purple in August or September. The pit has a distinct shape – keeled on one edge and flat with a groove on the other.

The Sierra plum *(Prunus subcordata)* is the largest stone fruit native to the Sierra and the most readily edible one. It is recommended as excellent for making jams and jellies. It apparently fruits much more successfully farther north in its range, which extends into Oregon.

The stands we have investigated in the Sierra grow on the west slope in the yellow-pine and digger-pine belts up to about 4500 feet. This life zone might well be termed the poison-oak belt or rattlesnake community, indicating further challenge in collecting those fruits. Plum trees tend to grow in somewhat moist areas and along streams, but sometimes we find them on dry slopes. In the northern Sierra (Plumas County), plums are also found on the east slope.

The plants grow as small trees or shrubs in size. Short branchlets take on the aspect of sharp thorns. The larger branches are grey, stiff, and crooked. The leaves are roundish

SIERRA PLUM

life size

and finely toothed. These are attractive-looking plants at any time during the growing season. But in the early spring, they are truly spectacular with their displays of white blossoms. Look for these showy white bushes in April and May. This is the best season for locating Sierra plum trees which you may wish to watch for fruit later on. The flowers are about an inch across with five white petals and numerous stamens.

We would be interested to know whether there are stands of Sierra plum in the central Sierra which are good fruit producers. We have watched quite a number of patches in this region and all have produced bumper crops of fungus.

WESTERN CHOKECHERRY

The western chokecherry *(Prunus virginiana* var. *demissa)* is another small tree or shrub with edible stone fruits. It can sometimes be found growing alongside Sierra plums, but the chokecherry is much more successful in its fruit production. It grows along streams, at the edges of meadows and in other moist places. It grows up to about 6000 feet in the Sierra but is most often seen at 3000 or 4000 feet.

As a tree, this plant grows to about fifteen feet in height. More often, you will see the chokecherry growing as a shrub less than ten feet high. The branchlets are not thorn-like. The leaves are toothed and acutely pointed. In the spring, the beautiful white flower spikes make a showy display.

The cherries ripen to a dark red in late August or September. They hang down in clusters at the ends of branches, and a substantial quantity can easily be gathered from a good area.

Ripe chokecherries can be eaten raw but are quite sour and puckery. Even when made into jam or tarts, some of the astringent quality remains. The best solution is to mix them with apples. The color and flavor of the mixture are excellent, and the apples seem to absorb the astringency. Conveniently, western chokecherries come into season at the same time as the apples in the many abandoned orchards of the same elevations in the Sierra.

WESTERN CHOKECHERRY

life size

Chokecherry-Apple Jam

Peel, core, and chop about 3 cups of apples. Cook them in a small amount of water until they are as soft as you want – either in tender chunks or as smooth applesauce. Remove the stones from 1 cup of chokecherries (sometimes it helps to boil them for a minute or two to loosen the stones). Mix the cherries and apples. Add about 2 cups of sugar and 1 tablespoon lemon juice and 2 teaspoons cinnamon. Boil the mixture until it thickens. Pour into sterilized jars and seal.

Chokecherries are not to be confused with bitter cherries *(Prunus emarginata)*. One taste of the latter will enable you to remember the common name without difficulty. The bitter cherry also grows as a shrub in the same range, but the leaves are thicker and less pointed, and the berries grow in many small bunches along the branches. You will soon notice how much more common and widespread the bitter cherry is than our edible stone fruits and how much more prolifically it fruits.

Chokecherries are a great favorite of many kinds of birds. In some areas, we have seen every last cherry go to the birds, most being eaten long before they ripened. In other areas, however, long clusters of ripe cherries can be found hanging in abundance.

HAZELNUTS

The California hazelnut or beaked filbert *(Corylus cornuta* var. *californica)* has the most civilized appearing wild nut of the Sierra, as it looks and tastes very much like the domestic filbert. But the wild hazelnut and the domestic filbert grow quite differently. The wild hazelnut is more of a shrub, produces fewer nuts and has a beaked sheath of bracts around the nut. Indians of the Sierra made limited use of these good nuts, which are too uncommon to have been a major food. Now our only competitors in collecting them seem to be squirrels, bears, birds, and worms.

California hazelnuts grow as small trees or shrubs in forested surroundings up to about 6000 feet on the west slope of the Sierra. The largest patches we have found have been in disrupted areas which have been logged and bulldozed. However, we've also seen them in primeval giant sequoia groves. (Come to think of it, these areas have probably been bulldozed, burned, and otherwise disrupted, too.)

The leaves of the hazelnut are toothed and somewhat fuzzy, resembling alder leaves. The key identification feature is the nuts with their beaked sheathes. The beaked sheathes have minute bristles which stick in your fingers in a most irritating manner. We generally wear gloves to pick hazelnuts and squeeze the nuts out of the sheathes. Often two and less often three of these nuts will grow around the same point near the end of a branch. They are likely to be hidden under the leaves and you may miss them unless you put your head down and

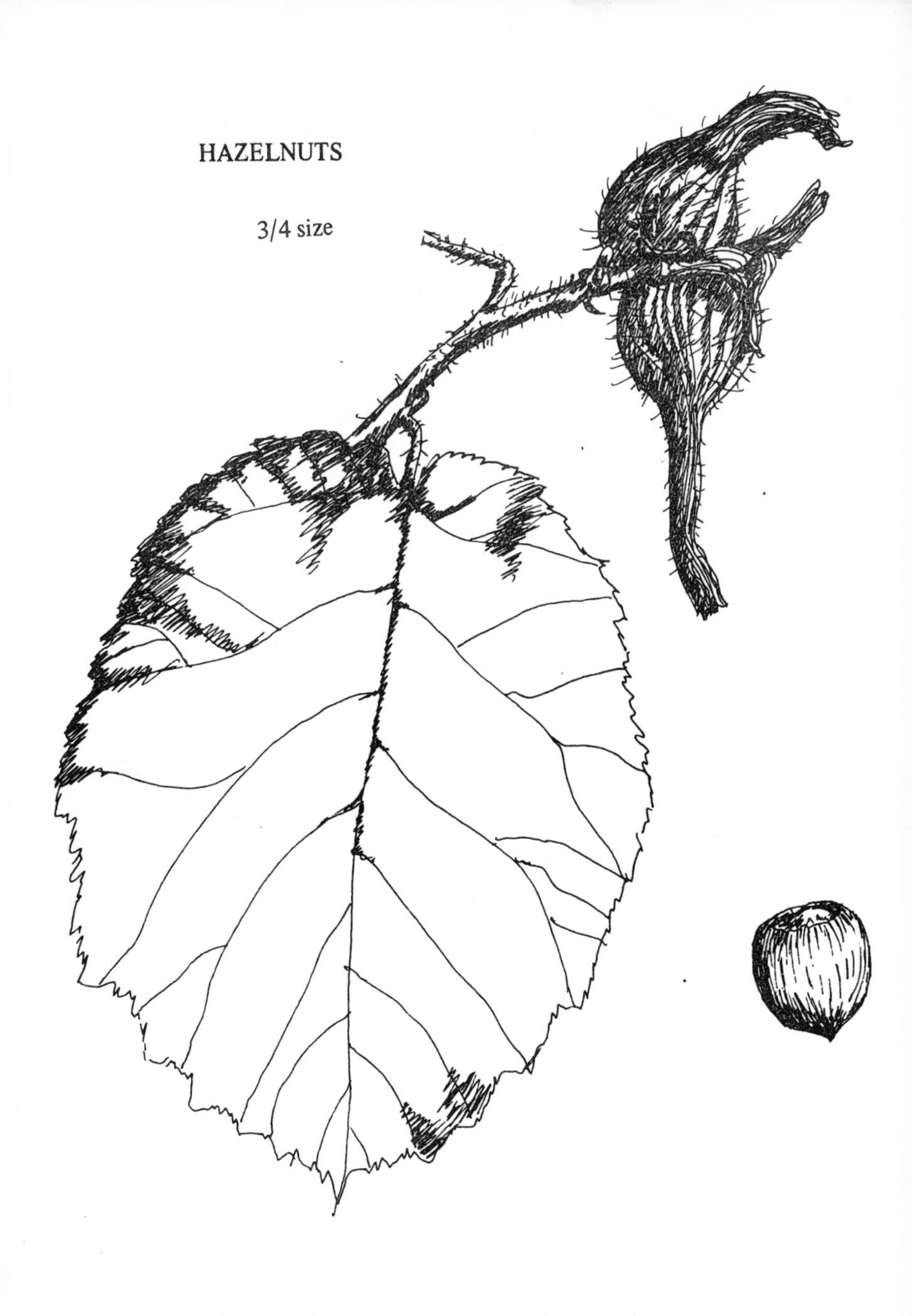
HAZELNUTS
3/4 size

look up. Many hazelnut bushes will be found producing no nuts at all for the season. You may often find bushes loaded with the male catkins but with no (female) nuts, but you should be able to find bushes producing nuts somewhere nearby.

The nuts can be seen developing as early as June or July. They are ready to pick in early September. If you are too early, the nutmeat will not have developed enough to fill the shell. If you are too late, you will see the remnants of the squirrels' feast. An especially maddening fact is that the squirrels can determine the hollow or rotten nuts without opening them. One year when we arrived a little late at a large patch of hazels, we laboriously scrounged a bagful of the few remaining nuts before we realized that nearly all were squirrel rejects.

Worm holes are conspicuous. You should leave those and let the creature continue its career.

We were especially impressed with a large bear scat we found in one hazel patch. The bear had apparently been eating mainly hazelnuts – shell and all, of course.

Wild hazelnuts taste fine when eaten fresh from the bush, although they are even better after drying. No processing of any sort is required. Just use them in any way you would use filberts.

SIERRA CHINQUAPIN

Lurking within that ball of spines on the Sierra chinquapin *(Castanopsis sempervirens)* is a tasty nut. The nut shell looks something like a small, pointed, three-sided hazelnut. The nutmeat is encased within a dry membrane which is easily peeled away. The nut is rich and sweet tasting, not at all bitter when mature.

Chinquapin nuts ripen in September to early October. At this time, you can see the points of the shells peeking out between spines. Each burr contains one to three nuts. Often, you will find just one of the nuts fully developed, while the other two are very small. Good nuts should be about one-third inch across, so keep looking if you are only finding small ones.

The sharp spines present a problem when collecting the nuts. I tried using thick leather gloves and was still stabbed right through the leather. One way to remove the nuts is to place the burr on a flat rock and roll the sole of your shoe over it. The nuts should come popping out – if you don't crush them in the process.

These evergreen bushes are very common in the Sierra between about 5000 and 10,000 feet. They grow in open forests and in clearings among other shrubs such as huckleberry oak and greenleaf manzanita. In July and August, Sierra chinquapins can be identified by the smell of the flower catkins, which leave you gasping for fresh air. Even a biologist friend of ours once mistook this smell for air pollution when he hiked past a chinquapin patch in the vicinity of a large

SIERRA CHINQUAPIN

life size

tungsten mill in the Sierra. Actually the chinquapin flowers do not smell nearly so bad as the sulphurous emissions which the mill blows far into the Sierra wilderness.

RED DOGWOOD AND RED DOGWOOD PARAPHERNALIA

Those of us who long ago read Horace Kephart's *Book of Camping and Woodcraft* will remember red osier dogwood as the frontiersman's tobacco substitute with that "narcotic" effect "approaching stupefaction." As a matter of fact, we do know of one fellow who, after smoking it steadily for several days while trying to give up cigarettes, reported some pretty bizarre effects. However, everyone with whom we have tested it has described only mildly pleasant, euphoric feelings. A few of the subjects reported no effect at all.

What is this plant and where does it grow? Red dogwood *(Cornus stolonifera)*, also called red osier dogwood and creek dogwood, is a beautiful shrub belonging to the same genus as the flowering dogwood tree. It is also referred to as "red willow" in the literature, but is not to be confused with *Salix*, the genus of true willows. The small white flowers grow in round-topped clusters. Later, these become berries which are avidly sought by birds. The thinner branches are reddish, especially the ones that reach into the sunlight. They become a brighter red-purple in the winter when the leaves are gone. Red-dogwood thickets are very common along streams on both sides of the Sierra up to about 8000 feet. The bushes are very attractive and are sometimes seen as plantings around buildings in the mountains.

To prepare the smoking substance, take a few of the reddest limbs you can find. It is even better to collect them in the winter when they are redder. Use a knifeblade to scrape off the thin red outer bark and discard it. The greenish inner bark is what you want. Hold the freshly scraped twig to your nose and smell the sweet fragrance resembling cantaloupe. Now hold it over a plate or piece of paper and scrape off the greenish inner bark. I find it easiest to work on one section between nodes at a time. Scraping off thin shavings in this manner is better than peeling off the whole layer of inner bark at once. Very fine shavings burn cooler. Avoid adulterating the product with the white wood under the inner bark. The wood is harder so with a light pressure the knife will stop shaving when the inner bark is gone. The inner bark shavings can either be left to dry or, if you lack patience, can be dried in a frying pan or oven at low heat. If you have no pan, try shredding the inner bark so that it clings to the twig in a frazzled condition and roast this over a fire.

If you don't have your own special smoking equipment and techniques, you can easily make a pipe of the style used by the Miwok Indians of this region. Cut a short, fairly thick section of elderberry stem. You should be able to find one that is already dead. Hollow out the pith with a wire. Insert a small piece of hardwood at the mouth end so that it mostly, but not entirely, blocks the passage. The dried bark is inserted in the other end, and the plug keeps it from being sucked into your mouth. You'll notice that you can use this pipe only in a reclining position, which would be appropriate anyway. (Note: Elderberry stems have been reported to be mildly poisonous to children. Probably fresh green stems were the offenders.)

We have tried red-dogwood bark only in small amounts. The smoke is very aromatic and rich tasting. The effects have been mild and include a visual "brightness" and sort of a warm,

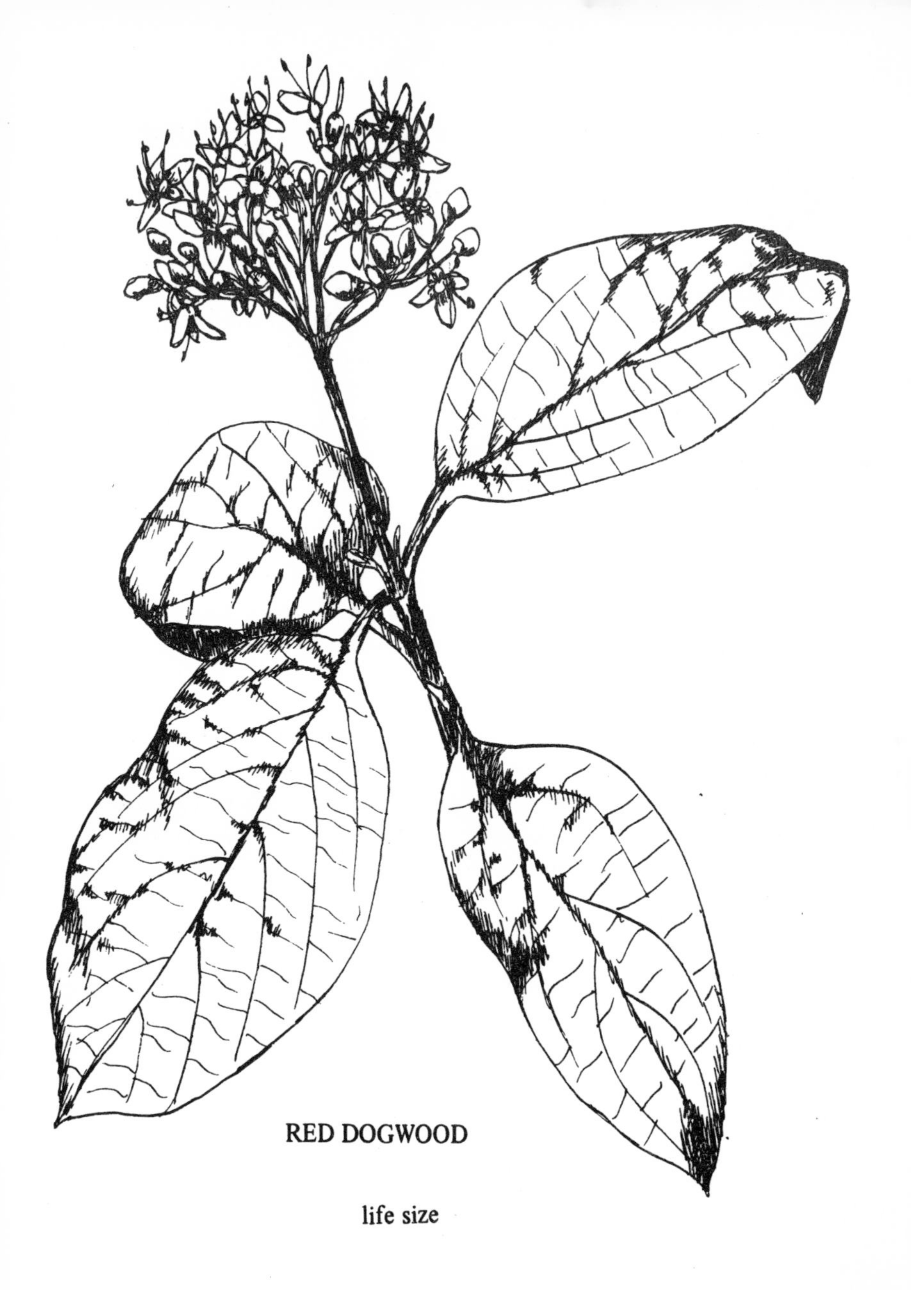

RED DOGWOOD

life size

serene feeling inside. The smoke made us feel rather at home in our mountain surroundings.

One time when we took some red willow twigs home our cat decided it was better than catnip and went wild. When she discovered the twigs on the table, she proceeded to chew them, wrestle with them, and throw them madly about. After an hour or so of this, I salvaged what I could and continued to collect the inner bark for my own purposes. Later, the cat discovered the sack of bark shavings in the cupboard, dragged it out and scattered the pieces about the house, dashing after them at incredible speed.

For a serious and much more authoritative treatment of the subject than either I or Kephart can give, I recommend *Black Elk Speaks* as told to John Neihardt. Black Elk, the Sioux holy man, would not even agree to tell Mr. Neihardt his story until the white man had smoked the "red willow" with him. Throughout the story, Black Elk uses the sacred smoke to help him invoke the Great Spirit.

Indians of the Sierra used much stronger medicine for this purpose, too. Datura *(Datura meteloides)*, known also as tolguacha and Jimson weed and seen producing its large, trumpet-shaped white flowers in dry, open areas of the foothills, was taken at significant times to induce visions and invoke spirits. Its principal use was in the initiation rites of young men and women when they reached puberty. After receiving a proper dose of datura, a youth wandered into the wilderness to spend a few days fasting and meditating until a vision came to him which helped establish his relation to his society and the rest of the universe.

That was fine for the Indians, who had developed a culture which related to such an experience and who had expert shamans to regulate the dosage. But we wouldn't touch this dangerous plant ourselves. Datura has fatally poisoned many humans, even Indians when all did not go just right. The

Indians had many generations of experience to rely on when using this enigmatic herb. They had a strong concept of "datura abuse" and frowned on anyone who used it excessively or for selfish purposes. It is doubtful that we could gain the same experience through these plants as did the Indians, who used them carefully in the proper context.

EPHEDRA

The wild "tea" that tastes the most like ordinary tea is ephedra *(Ephedra viridis).* Also known as Mormon tea, this primitive-looking bush is found in the piñon-juniper woodlands of the east side of the Sierra up to about 7000 feet. The yellow-green stems are jointed and the leaves are small scale-like structures which hardly look like leaves at all. Look for these strange plants in dry, rocky areas.

To make ephedra tea, add a small handful of twigs to a quart of boiling water and steep until the desired strength is obtained. You may want to simmer the twigs for a few minutes to bring out more flavor. Ephedra tea does not become bitter when boiled, and it is slightly sweeter than ordinary tea.

Ephedra tea was taken for colds and stomach aches by the Indians and early settlers. It has also been alleged to produce a stimulative effect, but if this is so, we have not noticed it and it must be mild.

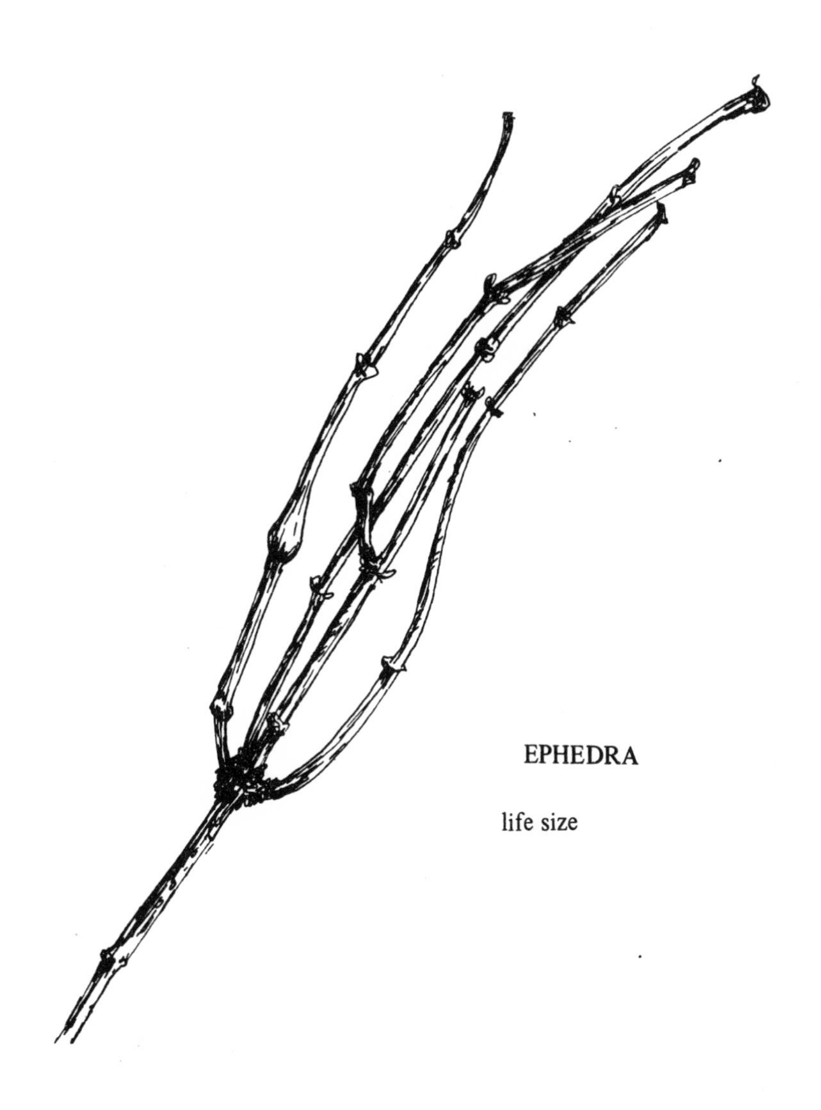

EPHEDRA

life size

YERBA SANTA

Yerba santa *(Eriodictyon californicum)* received the name "holy herb" from the Spanish Californians in recognition of its healing and health-giving qualities. The Indians taught the early settlers to use this plant for colds, coughs, stomach ache, asthma, and other ailments. We enjoy sipping yerba santa tea simply for enjoyment. Two or three leaves can be torn up and steeped to make a cup of sweet, aromatic beverage which is never bitter, even when boiled. The leaves can be dried for future tea making. Select the leaves with care, taking only healthy-looking ones, and do not pick many from any one bush.

This unique shrub grows in chaparral and open forests and on dry, sunny slopes of the Sierra foothills and mixed-conifer belt up to about 5000 feet. It is especially prevalent on sunny disrupted areas such as roadsides. Yerba santa can be recognized by the thick, leathery nature of its brownish-green leaves. The leaves look sticky but usually do not feel sticky. Deer like to eat the chewy leaves, and you will often see the signs of their browsing. The small flowers of yerba santa are white to lavender.

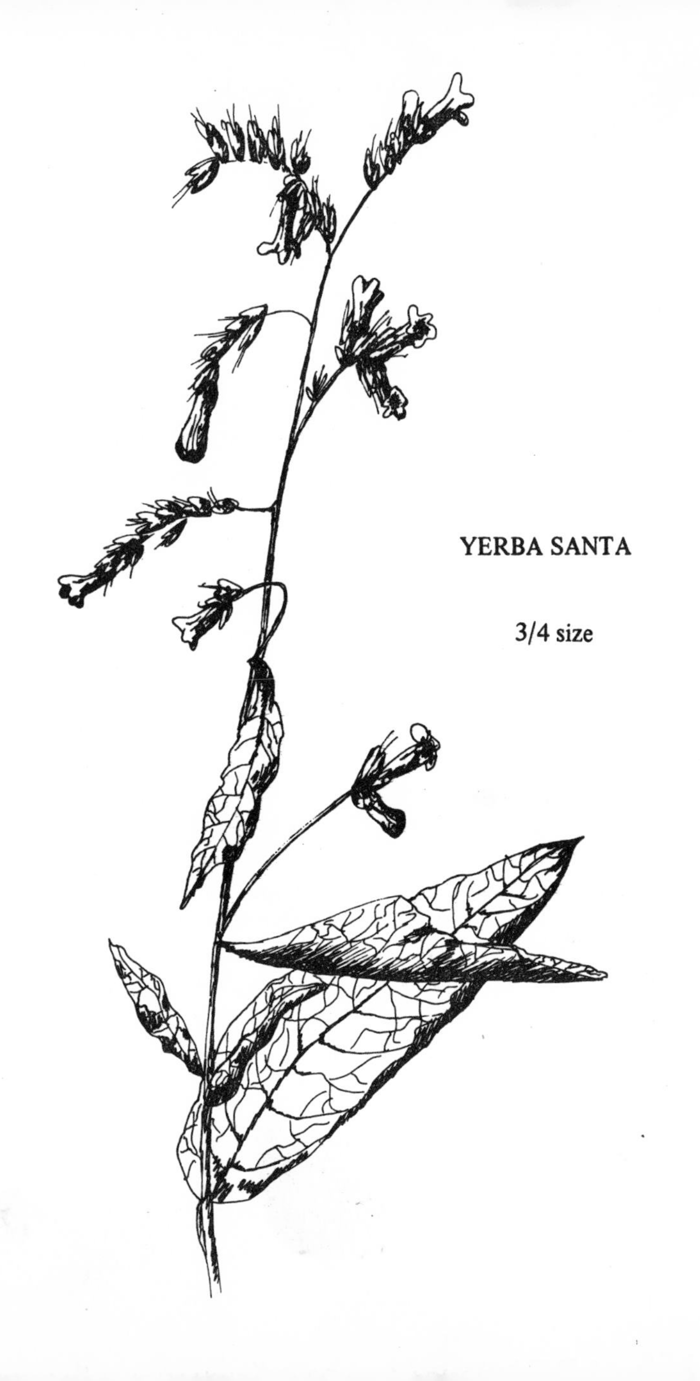

YERBA SANTA

3/4 size

WILD GRAPES

John Muir wrote in praise of a certain natural vineyard which grew in a beautiful secluded spot in the Sierra. The same vineyard still flourishes. It is one of our favorite places, but it produces hardly any grapes.

California wild grapevines *(Vitis californica)* grow thickly along canyonsides of the west slope of the Sierra up to about 4500 feet. Unfortunately, John Muir's vineyard is typical of many of them, especially the ones at higher elevations. Some vines can be found, though, that are heavily laden with hanging clusters of wild grapes. One such area is all you need to find. Grapevines grow along the ground and up over rocks and trees. Some of the vines become several inches in diameter and can be climbed to reach high clusters.

Greenish-yellow blossoms can be seen in the spring and early summer. The grapes are slow to mature in the mountains, and most of the ripe grapes will be found in September or later. California wild grapes grow to about a third or a half inch in diameter and are purple when ripe. Fully-ripe wild grapes are quite good to eat raw. Large pear-shaped seeds will be found inside.

Mary makes an excellent jelly from the ripe wild grapes, which seem to have plenty of natural pectin.

Wild Grape Jelly

Use about 1/4 underripe and 3/4 fully ripe grapes. Remove the grapes from the stem, crush them, add a little

WILD GRAPES

3/4 size

water (1/2 cup) and bring to boil. Simmer for about ten minutes. Extract the juice using a jelly bag or some cheesecloth.

Use 4 cups juice to 3 cups sugar. Stir sugar in well at low heat until sugar is dissolved. Then rapidly heat the mixture to 8 degrees Farenheit above the boiling point of water. (Remember the boiling point varies with altitude.) If you have no thermometer, heat the mixture until it sheets from a spoon. (As the mixture heats it becomes thicker. Take a small amount and pour it off the edge of a spoon. At first it will drip one drop at a time. As it becomes thicker it drips two drops side by side. Next it comes off in a sheet. This is the point at which to take it off the fire.) Pour into sterilized jars and seal immediately.

Mary's equipment for jelly making in the mountains consists of one 2-quart saucepan, pint canning jars, and a No. 10 tin can. Two jars and their lids can be sterilized at one time by bringing them to a boil in the can. Sealing can be done with paraffin instead of lids, but that requires an extra pot for melting the paraffin. If you are camping in the Sierra and don't have as much as a can or jars, you can always make a small amount of wild jelly in a single pot to taste but not to save. These wild grapes are not well suited for jam as their skins are quite tough.

Earlier in the summer unripe grapes can be used to make tasty jelly. Mary made green grape jelly in mid-July with grapes from a vine which didn't have ripe fruit until the middle of September. Make it the same way as purple grape jelly, only adjust the sugar if the grapes are very sour. Try 4 cups juice to 3 1/2 cups sugar. The jelly will be quite tangy anyway.

ELDERBERRIES

One kind of wild berry that can easily be gathered in great quantity in the Sierra is the elderberry. The best species of elderberry for eating is the mountain blue elderberry *(Sambucus caerulea)*. These tall shrubs are common on both sides of the Sierra up to about 8000 feet. They grow along streams, on dry slopes and flats, and are especially common is disrupted areas such as roadsides, logged and burned areas, and mining wastelands. The thickest patch we ever found was at 5000 feet in a wide swath that had been cleared for power lines.

This species of elderberry usually has seven leaflets to each leaf. The flowers are white and grow in clusters with flat tops. These develop into dark-blue berries with a whitish bloom. Ripe berries can be found from late July to early October.

I like to eat the fully ripe berries raw, but not everyone shares my taste for tart fruit. These berries make fine jam, jelly, syrup, pies, tarts, and crisp. The color and flavor are both intense. The berries are somewhat seedy, but the seeds are small and soft.

MOUNTAIN BLUE ELDERBERRY

3/4 size

One-Pot Elderberry Crisp

Crisp part:

1/4 cup butter

1 cup oatmeal

1/2 cup wheat germ

1/2 cup brown sugar

Melt butter and stir in everything else. Cook for a few minutes until it's sort of crisp. Put aside in a cup or plate.

Elderberry part:

1 quart elderberries

2 cups sugar (or to taste)

1 teaspoon lemon juice

2 tablespoons cornstarch that has been dissolved in a little water

2 teaspoons cinnamon

Cook until berries burst and mixture is thickened. Put the crisp part on top and eat.

Elderberry Jelly and Jam

Put berries in a pot with a little water and boil until they burst. Strain the juice with cheesecloth or a jelly bag. Mix these proportions: 4 cups juice; 3 cups sugar; 2 tablespoons lemon juice. Bring slowly to a boil while sugar is dissolving and then cook rapidly to jellying point (refer to Wild Grape Jelly). Pour into sterilized jars and seal.

To make jam put berries, sugar, and lemon juice in a pot all at once and proceed with the cooking. The jamming point is slightly different – 9 degrees above the boiling point of water or until it rounds from a spoon (i.e., a spoonful is thick enough to stay rounded on top).

Another blue elderberry, *Sambucus mexicana,* grows at lower elevations with three to five leaflets per leaf and smaller leaflets and flower clusters. We have found the berries of this species inferior to the mountain variety. They tend to be less juicy, seedier, and have a bitter taste which does not disappear with cooking or sweetening.

The mountain red elderberry *(S. microbotrys)* is common at higher elevations in the Sierra. However, the red berries are said to cause gastrointestinal upsets in some people, so it is probably wise to avoid them.

Having heard much about "delicious elderblossom fritters" we tried making them with the blossoms of our mountain blue elderberry. We cut a fresh cyme of blossoms, dipped it in batter and fried it to a golden crisp. Ugh. It was about as bitter as anything could be. Perhaps this depends on the plant. Other parts of the elderberry plant are reported to be mildly poisonous.

The stems of elderberry are woody with soft pith inside. The Miwoks made flutes and various other musical instruments out of hollowed-out elderberry shoots. For detailed instructions on making an elderberry-stem pipe, refer to the section on red dogwood. Remember to cut only stems which are already dead.

BLACKBERRIES, RASPBERRIES, AND THIMBLEBERRIES

Three species of the genus *Rubus* provide us with some good berries in the Sierra.

The California blackberry *(Rubus vitifolius)* can be found around the 4000 foot level and lower in dampish woods. The most extensive briar patches of this species we have found grow at about 3000 feet. The berries are much like commercial blackberries except smaller and tastier. The oblong black berries are half an inch or less in length.

The largest, juiciest berries always seem to grow deepest within the impenetrable thicket, but you can still collect a large amount from a good patch. The berries ripen throughout July, August, and September. It is easy to eat them as fast as you pick them, but if you have some restraint, they are fine for pies, jams, and jellies. When cooked in pies or jams, for some reason the berries seem seedier than they do raw. But the product is still good.

Escaped domestic blackberries are also found at these lower elevations, especially around old orchards.

Another thorny species with good berries is the western raspberry *(Rubus leucodermis)*. It can be found at higher elevations, up to about 7000 feet in the Sierra. We most often come upon these bushes in steep canyons and near streams. The berries are a dark purple when ripe in July and August. When you pick one, it comes loose in sort of a hollow cone shape. Raspberries make good jellies and somewhat seedy but good jams. It is hard to find enough at once in the Sierra for pies.

Wild Blackberry or Raspberry Jam

Use 4 cups berries to 2 cups sugar. Bring slowly to boil, stirring until sugar dissolves. Then cook more rapidly to the jamming point – 9 degrees above the boiling point of water or until it rounds up in a spoon (that is, a spoonful is thick enough to stay rounded on top). Pour hot into hot, sterilized jars and seal.

Our favorite berry of this genus is the thimbleberry *(Rubus parviflorus)*. It grows to higher elevations, up to 8000 feet on both sides of the Sierra. Thimbleberries grow in damp woods and canyons and near streams. The leaves are large and shaped like maple leaves, and there are no thorns. The large white flowers are followed by round scarlet fruits which are delicious in the Sierra. I say in the Sierra because we have eaten thimbleberries in the Coast Ranges which tasted mediocre at best.

One of my favorite fantasies is thimbleberry preserves. I have never been able to live this one out, however, because we have never been able to pick these berries faster than we eat them.

The fresh leaves of western raspberry and California blackberry can be brewed into good, healthful tea. Boiling does not hurt the flavor but may lower the vitamin content. Dried leaves can also be used for tea. We have never tried making tea with thimbleberry leaves. Perhaps the somewhat hairy nature of these leaves makes them less enticing for this use.

THIMBLEBERRY

life size

GOOSEBERRIES AND CURRANTS

There are about a dozen species of wild currants and gooseberries in the Sierra. Loosely speaking, when a shrub of the genus *Ribes* has thorns on the branches it is called a gooseberry, while those without thorns are referred to as currants. The gooseberry fruit may be either spined or unarmed. There is a great deal of difference in appearance and flavor among different species of gooseberries and currants.

One of the first species you are likely to notice in the Sierra is the Sierra Gooseberry *(Ribes Roezlii)*. The berry is half an inch or more in diameter and is covered with stiff, sharp spines. It is bright red-purple when ripe. The thorny bushes are found in forests and open slopes of the mixed-coniferous and red-fir forests between 3500 and 8500 feet. We have found the thickest stands of these in logged-over areas which have been bulldozed and otherwise disrupted. The berries usually ripen in August though some can be found as early as July or as late as September.

Gloves are very helpful in picking the spiny berries. The fresh raw berries have a very good taste, but because of the spines and thick skins, they are best suited for making jelly. They contain enough natural pectin for this purpose, and the flavor and color of the jelly are superb.

Sierra Gooseberry Jelly

Put berries in a pot with a little water and simmer until soft. Put the cooked berries in a jelly bag or square of cheesecloth and press the juice through. Mix 4 cups juice, 2 cups sugar and one tablespoon lemon juice. Boil the mixture until it reaches the jellying point – 8 degrees above the boiling point of water in your locality, or until it sheets from a spoon. (Refer to Wild Grape Jelly for further explanation). Pour into sterilized jars and seal immediately.

SIERRA GOOSEBERRY

life size

Another common *Ribes* you will run into in the same general area is the Sierra Currant *(Ribes nevadense)*. This shrub is found in moist places, mainly along streams. The leaves look something like small maple leaves, two to three inches wide. The berries are blue-black, often with a whitish covering on the surface called a bloom. These wild currants can be eaten raw but are much better when made into jelly. They are a bit seedy for jam. The jelly is very tasty and it turns out a much more impressive color than the raw berries.

Sierra Currant Jelly

Put the currants into a pot with a little water and simmer until soft and juicy. Press through a jelly bag, cloth, or strainer. Add 2 cups sugar to 4 cups juice. Stir in sugar slowly until dissolved. At this point add 1 tablespoon lemon juice and presto – the berry juice turns a beautiful red-purple. Maintain a rolling boil until it reaches the jellying point. Pour into sterilized jars and seal.

A poisonous berry to watch out for, which shares the moist habitat of the Sierra currant, is the baneberry *(Actaea rubra* ssp. *arguta)*. The berries are either red or white. In the Sierra, we have only seen red ones. The berries grow on a stalk at the top of the plant.

Our two favorite *Ribes* species grow at high elevations, which suits us since we always spend a good part of the summer in the Sierra high country. The alpine prickly currant *(Ribes montigenum)* grows between 7000 and 12,000 feet in open, rocky places and under whitebark pines, lodgepole pines, and red firs. The "prickles" on the bright-red berries have merely been soft hairs in our experience, and the berries can be eaten raw, "prickles" and all. The thorns on the bush are not soft, however, so the alpine prickly currant might better be called a gooseberry. The alpine gooseberry *(Ribes*

SIERRA CURRANT

life size

lasianthum) grows between 7000 and 10,000 feet in the same habitats as the alpine prickly currant. The alpine gooseberry bushes have thorns but the bright-red berries are smooth.

While the berries of both species can be eaten raw, they are much better cooked. They are excellent for pies, cobblers, crisps, jams, and jellies. Crisps, unlike pies and cobblers, are easily made on a campfire or mountaineering stove. Follow the recipe for elderberry crisp using four cups of berries and one to one and three-fourths cups of sugar to taste. But jelly is our favorite use for alpine gooseberries and prickly currants. It is a beautiful scarlet color and is very tasty. Just follow the recipe for Sierra gooseberry jelly.

The squaw currant *(Ribes cereum)* has a round, red berry that looks almost exactly like the alpine gooseberry and the alpine prickly currant. The squaw currant shares the dry, rocky habitat of these other two species and is sometimes found growing side by side with them. The squaw currant does not have any thorns on the branches or on the berries. The berries would probably make a good jelly, but they are not as juicy nor as tasty as either alpine prickly currants or alpine gooseberries. In appearance, the leaves and bushes of these three species are fairly similar to those of the Sierra gooseberry.

Gooseberries and currants are among the few wild fruits of the Sierra that can be gathered in quantity. You may wish to experiment with some of the other species of *Ribes* found in the Sierra.

Gooseberries and currants have long been the object of extermination programs conducted by the Forest Service and even the Park Service in the Sierra. This unfortunate persecution results from *Ribes* being the intermediate host of white-pine blister rust, an exotic parasite which attacks five-needle pines such as the whitebark pine and the commercially valuable sugar pine. It seems ironic that species of *Ribes* are among the first plants to invade an area that has been disrupted by logging operations.

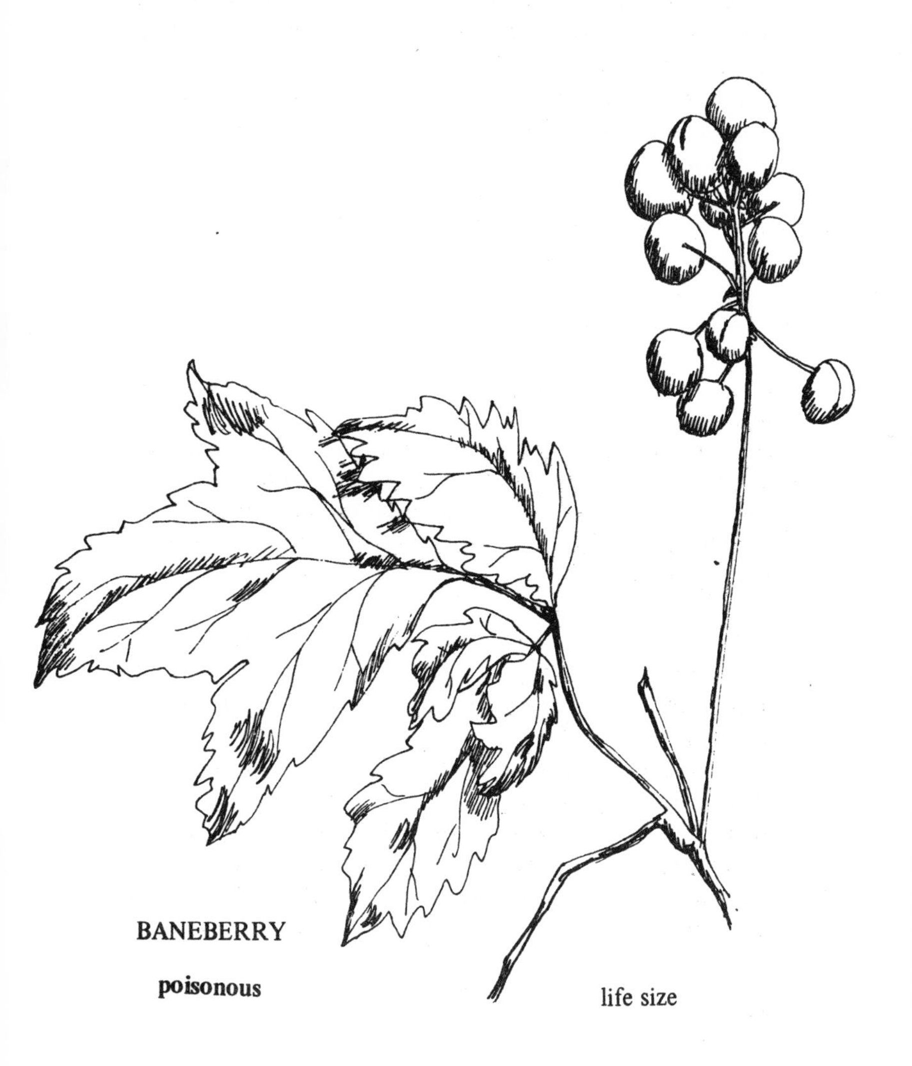

MANZANITA

The serious bushwhacker finds his ultimate challenge in the dense stands of manzanita *(Arctostaphylos* species) of the Sierra. As he emerges from a long traverse of a spectacular thicket, the shreds of his trousers and shirt hang in mute testimony to his Victory over Nature. Many techniques have been utilized against the unyielding, crooked branches of this shrub. One method is to walk over the tops of the bushes, balancing on the branches. After falling through, you may try wriggling beneath the branches on your stomach. Stuck? Often an appropriate expletive is effective when all else has failed. The real secret to the art of manzanita bushwhacking, however, is to lose your cool and thrash.

There are several species of manzanita in the Sierra. All have more or less edible berries. The berries ripen in August and September. They can still be used, however, even later after they have dried on the bushes. The tough, leathery, evergreen leaves and the smooth, reddish bark make most species easy to recognize.

The greenleaf manzanita *(A. patula)* is the dominant type of manzanita throughout the coniferous forests of the Sierra. It is common between 4000 and 8000 feet in pine and fir forests, rocky places, along roadsides, and in clear areas, where it grows in thick stands. The leaves are roundish and bright green. The bushes grow three to six feet in height. The several stems sprout from a swollen base although this is often buried under the humus.

GREENLEAF MANZANITA

life size

The berries of greenleaf manzanita become light red-brown to blackish when ripe. They are rather dry and seedy but very sour and can be used to make a flavorful beverage. We simply boil the berries in water for about fifteen minutes, mashing them as they get soft so all the flavor is extracted. This makes a very sour drink, but when sweetened with sugar or honey, it resembles hot lemonade.

The Miwoks had a more refined method for making manzanita cider, which probably resulted in a more nourishing drink than ours. They ground up the berries, placed the pulp in a strainer basket and poured hot water repeatedly through it until the cider was strong enough. This method should preserve the vitamins better as the drink is not actually boiled.

We have steeped the pink blossoms of greenleaf manzanita to make a delicious tea.

We have also tried making hot cider with the sticky red berries of whiteleaf manzanita *(A. viscida).* This species is common at lower elevations, up to about 5000 feet. The bushes grow very tall, to about twelve feet, and the leaves are whitish green. The sticky berries are unappealing to handle and the beverage proved quite flavorless.

The manzanita which grows at the highest elevations in the Sierra is pinemat manzanita *(A. nevadensis).* This low, sprawling shrub looks like little more than a bright green ground cover. It is fairly common between 5000 and 10,000 feet in forests and out among the rocks. The orange-brown berries make a good cider. The flavor has both sweet and sour aspects and does not need to be sweetened further.

Manzanitas are perhaps the most conspicuous and widespread shrubs in the Sierra. They are often seen growing along with other shrubs, particularly huckleberry oak and chinquapin. "Huckleberry oak and chinquapin!" scoffs one of our top-rated bushwhackers. "I could do them in shorts!"

PINEMAT MANZANITA

life size

BLUEBERRIES

Blueberry bushes are common in the high country of the Sierra. The berries are sweet and tasty. But ripe berries are found growing only sparsely on scattered plants. These small berries are usually inconspicuous among the leaves. Hence, blueberries are usually overlooked as edible berries in the Sierra.

The western blueberry *(Vaccinium occidentale)* grows as a low shrub. You will find these attractive, whitish-green bushes growing around wet meadows and along streams from 5000 to 11,000 feet or higher. In June and early July, the white to pinkish bell-shaped flowers can be seen if you look closely. The berries ripen throughout August and September, depending on the exposure. They are oblong, about one-fourth inch in length and are blue-black with a whitish bloom. The berries seem to grow somewhat larger at lower elevation.

When fully ripe, these berries can be very juicy and good. If they are even the slightest bit underripe, however, they are much less tasty. It is virtually impossible to gather more than you would want to eat on the spot, so it is pointless to consider using them in cooking. You will come upon many western blueberry bushes without any fruit on them. In another area, you might find several bushes together producing a relatively good crop of berries.

Another species of blueberry that is even less noticed is the Sierra bilberry *(Vaccinium nivictum),* also referred to as dwarf blueberry. The leaves look like scarcely more than a ground

WESTERN BLUEBERRY

life size

cover on wet meadows from 7000 to 12,000 feet. They also form an elegant fringe around the edges of rocks in these meadows. In some areas, the little bell-shaped flowers make a subtle display of pink across the ground. Other whole patches do not bloom at all in some years. The berries, often hidden between the leaves, ripen blue-black with a whitish bloom, and are round. They become ripe in August and September. The flavor of Sierra bilberries is perhaps the better of the two species. They are very juicy and taste like concentrated domestic blueberries.

SIERRA BILBERRY

life size

In September and October, whole mountain meadows in the high Sierra turn a beautiful reddish color as the Sierra bilberries assume their fall colors. Western blueberry bushes also become conspicuously beautiful in autumn. This is a good time to spot blueberry bushes for next year's berries. Mountain meadows with both the soft mat of Sierra bilberry and the delicate trim of western blueberry bushes are perhaps the most park-like places in the Sierra.

You are in for an unpleasant shock if you mistake a twinberry *(Lonicera involucrata)* for a blueberry. These black-colored berries grow in pairs on bushes along streams. Having heard conflicting reports on their edibility, I once decided to try one. A tiny taste left me spitting and rinsing my mouth out with water for a good long time before the bitter sensation faded.

WILD ROSE

Many species of wild rose (*Rosa* species) grow in the Sierra. The prickly bushes can be found in many habitats such as dry slopes, moist places, along streams, and in various kinds of forests at all elevations up to about 11,000 feet. The rose-pink flowers make an attractive display from June to August.

Rose hips are one of our very favorite wild foods and one of the few that we collect in quantity each year. The hips are the bright red fruit of the rose plant. Rose hips are renowned for their high Vitamin C content, many times higher than that of oranges. Health-food stores often sell rose hip concoctions at high prices.

The size, shape, and flavor of rose hips varies from species to species and from bush to bush. Test one from an area and give preference to those that are tasty and sour. We like to gather the good, medium-sized rose hips from *Rosa Woodsii* at the same time that we collect piñon nuts in September. Great patches of this species with profuse amounts of the red fruits are found along streams in the piñon woodlands of the eastern side of the Sierra. It takes us about an hour or two to collect a year's supply of these vitamin-filled fruits.

Rose hips are very seedy and not very satisfactory to eat raw. But they make a delicious fruity tea. The hips should be chopped up and boiled or steeped to the desired strength. Sweetened with honey, the tangy tea becomes more like a hot punch.

Rose hips also make a tasty jelly. They are low in natural pectin so you must add commercial pectin or mix them with a high-pectin fruit such as apples.

Rose Hip-Apple Jam

Remove seeds and blossom ends of the rose hips and cut them up. Mix 1 cup cut-up rose hips with 2 cups peeled, cored and chopped apples. Boil the fruit with a little water until soft. Add 1 to 1 1/2 cups sugar (to taste) and 2 teaspoons cinnamon. Continue cooking until mixture is thickened. Pour into sterilized jars and seal.

If removing the seeds from many rose hips is too odious a task, merely make a cup of very, very strong "tea," strain it, and add it to 2 cups cooked apples. Then proceed as above.

Rose hips can be dried and stored for the rest of the year without losing any of their flavor as far as we can tell. In fact they smell much better after drying for awhile. Just let them dry slowly in the sun or dry air. Even those hips which have dried on the bush are good to use when you find them in the winter or spring. For some reason, the birds and squirrels do not devour rose hips at the rate that they eat many other wild fruits. Thus, we can always find a few for ourselves in almost any season.

The leaves of the rose bush also make a good tea. They can be used either fresh or dried. Pour boiling water over a small amount of the leaves and steep for a few minutes.

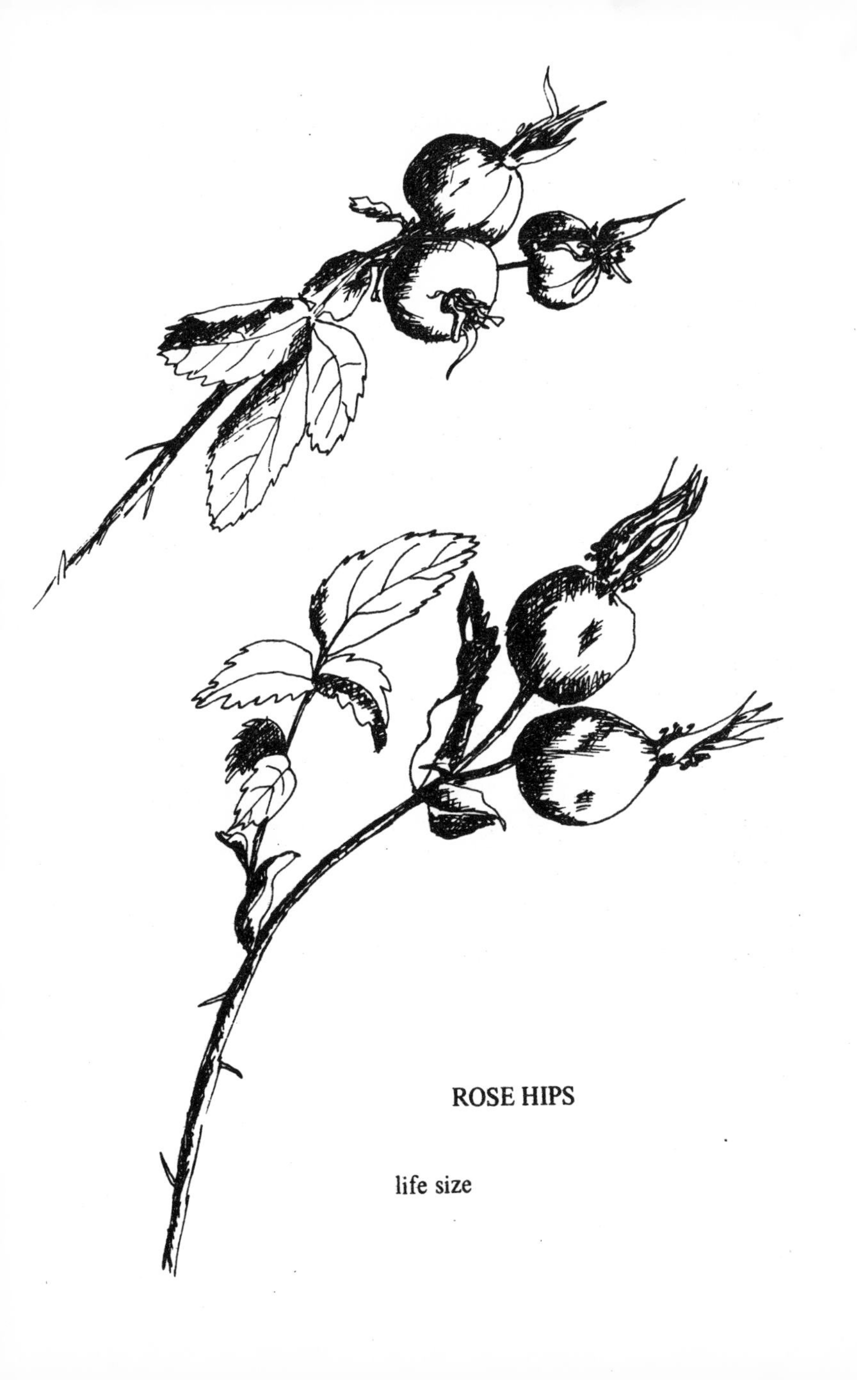

ROSE HIPS

life size

SERVICEBERRIES

We have never found serviceberries *(Amelanchier pallida)* growing as profusely in the Sierra as in various other ranges of the West. These decorative bushes are quite widespread here though. We find them growing on dry slopes and the sunny sides of canyons, under forest trees, and along streams. We have seen them from 4000 feet up to about 9000 feet in the Sierra.

The leaves have sharply-toothed edges. In spring and early summer, the bushes are lavishly covered with the many white blossoms. The berries ripen to a purple-black color in June, July, and August. They are usually about one-fourth inch in diameter, but on some bushes we have found berries considerably larger and juicier than the average.

Although the berries are sweet and juicy, they are not especially flavorful. Still, they make a refreshing fruit snack when found on a long hike. Serviceberries were often a component of pemmican, which Indians in various parts of North America carried for sustenance on long trips.

SERVICEBERRY

life size

WILD STRAWBERRIES

The wild strawberry is another plant which is very stingy in its fruit production in the Sierra. There are two species here, *Fragaria californica,* which grows below 7000 feet, and *F. platypetala,* which grows up to 10,500 feet. The two species are very similar in most characteristics, including the above.

Although extensive beds of strawberry plants are not uncommon in the Sierra, and often you will see quite a few of the white blossoms in the spring and early summer, finding the ripe red berries in August is a different matter. Many large patches will be found not producing a single fruit. Fortunately, you can occasionally find a patch with a fair number of fruits. We have found such patches of *F. platypetala* at 10,000 feet both east and west of the Sierra crest. They were growing on shaded, damp ground under lodgepole pines. The berries hide under the leaves. The flavor is so intense and concentrated that you will never again be satisfied with a cultivated strawberry.

Fresh strawberry leaves make one of the better wild teas. It is reputed to be very high in vitamin C. Pour boiling water over the leaves and steep. Dried strawberry leaves also make a good-tasting tea, though probably with less vitamin content. We have often seen dried strawberry-leaf tea for sale in health-food stores.

WILD STRAWBERRY

life size

CAT–TAIL

I would not initially think of the cat-tail *(Typha domingensis)* as a mountain plant. Yet we have seen cat-tails growing on both sides of the Sierra up to about 6000 feet. And cat-tails provide us with such a variety of good foods that it would be hard to ignore them as an edible species. Cat-tails grow in thick patches in swampy areas. Red-winged blackbirds often make their nests in them.

Our favorite cat-tail product is the pollen from the male flower spike, the top spike. When these flowers are bright yellow in June and July, the slightest tap will send great clouds of the yellow pollen into the air. By holding a plastic bag over the bent-over spikes and shaking them, you can collect the pollen. Be careful not to break the stem so the female flowers can develop later. Sift the pollen while you are still in the cat-tail patch so you can liberate the little bugs in their natural habitat.

The pollen can be mixed with flour and used in many recipes. The result is an appetizing color and a surprisingly delicious taste. One of Mary's campfire recipes with this pollen flour follows.

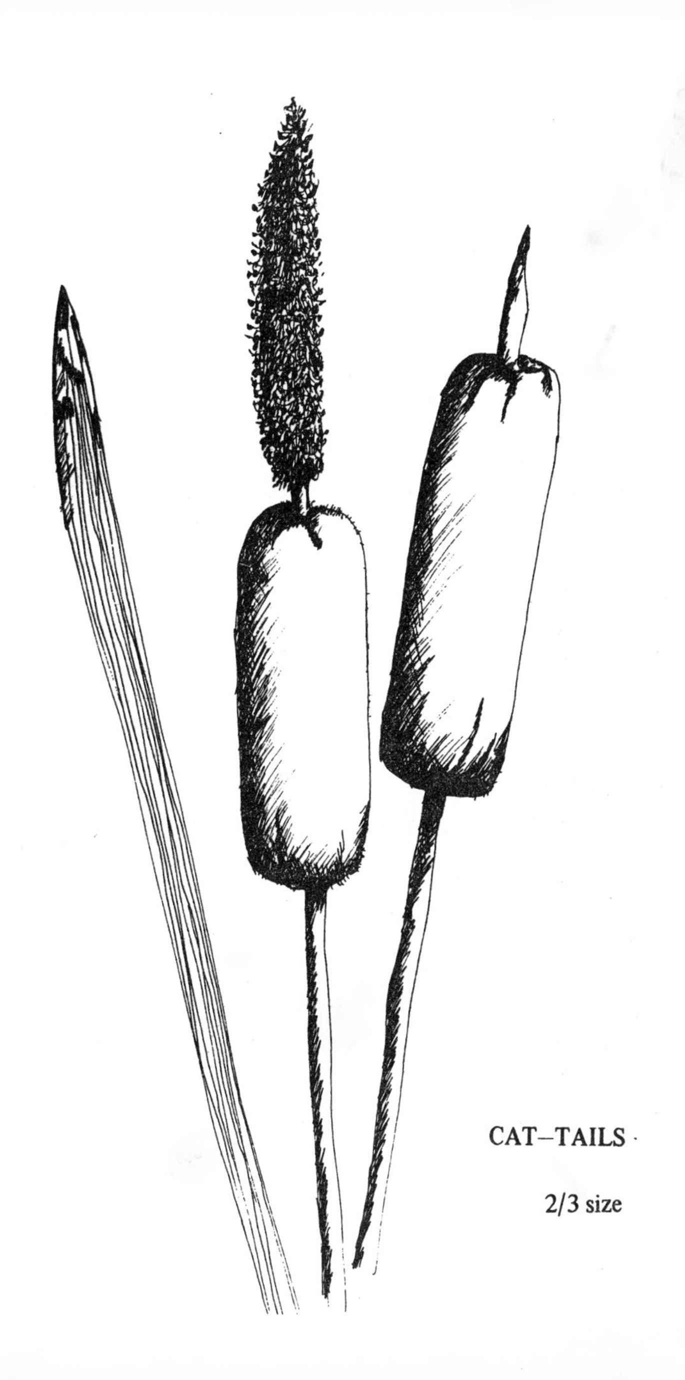

CAT–TAILS

2/3 size

Cat-tail Biscuits

Sift together:
1 cup whole wheat flour
1 cup sifted cat-tail pollen
2 teaspoons baking powder
1 teaspoon salt
Mix in:
1/4 cup vegetable oil
2/3 cup milk

Stir vigorously for about a minute. Drop by spoonfuls into your well-greased pan and carefully cook with the lid on over the fire or stove. You may want to turn the biscuits.

If you have an oven you can do a better job by baking the biscuits at 425 degrees for 10 to 12 minutes. You may want to roll out the dough and cut it into rounds.

If you have access to an oven, cat-tail pollen also makes a good addition to bread dough, oatmeal cookies, and corn bread. Replace one-half of the flour in the recipe with pollen.

Before the top spike blooms, it makes an extremely good wild vegetable. Cut off both spikes when they are still contained within a sheath or when the sheath is just coming off. Treat the spikes like ears of corn on the cob. Boil them for about five minutes or until the outer part is tender. These can be buttered and salted, and they taste something like corn. Both spikes can be eaten down to the "cob," but the lower spike has very little fleshy surface on it. The upper spike is covered with a fairly thick layer of delicious would-be blossoms.

A refreshing salad vegetable comes from the young cat-tail shoots. Grasp the plant at its base and pull it from the ground. It should break at just the right point. Young cat-tail plants up to about a foot and a half high are suitable for this use. Peel the shoots to expose the tender inner portion. These can be eaten either raw or cooked. We like them best raw. They are crunchy and very good tasting, reminding us somewhat of celery. It is possible to find these young shoots as late as July, when other cat-tails nearby are already producing pollen.

Some parts of cat-tail roots can be collected and eaten at any time of year. The rootstock just below the ground surface can be peeled and eaten raw but is best cooked. It is white and starchy in the center and has a good, though mild, taste. It seems obvious to compare it with potatoes. The inner portion of the thick horizontal rootstocks can also be eaten. It is tough, however, and even after cooking you can only chew out the starch and spit out the fibers. New sprouts growing from these horizontal roots are very tender and good tasting when peeled and boiled for about three minutes. They are even better when fried lightly. They make a very good vegetable served with butter and salt. Let's not forget the white starchy connection between the sprout and the center of the horizontal root. Cut it out and it is good and tender like the rootstock center.

If you find all this information about the roots confusing, simply look at it this way. Any of the white inner parts of the cat-tail roots are good to eat, though some portions are more tender than others.

A particularly nice characteristic of cat-tails is the way different shoots in a patch develop at different times during the growing season. Thus, one day in early July at 6000 feet we were able to collect young shoots, spikes, pollen, sprouts, and roots all from one cat-tail patch.

BULRUSH

The bulrush *(Scirpus acutus)* or common tule is a wild food plant similar to the cat-tail. Like the cat-tail it is found in thick patches in swampy areas, usually below 5000 feet.

The food uses are also quite similar. The pollen may be gathered in the early summer and used in recipes with flour. The young shoots may be peeled and eaten raw. The rootstocks have an edible starchy core, but it seems quite a bit tougher than the corresponding part from cat-tails. The horizontal roots of bulrushes have proven too tough and woody for us even to dig up. The best underground parts are the new sprouts. They are quite good either raw or boiled after being peeled. In general, however, we prefer the quality of the foods from cat-tails.

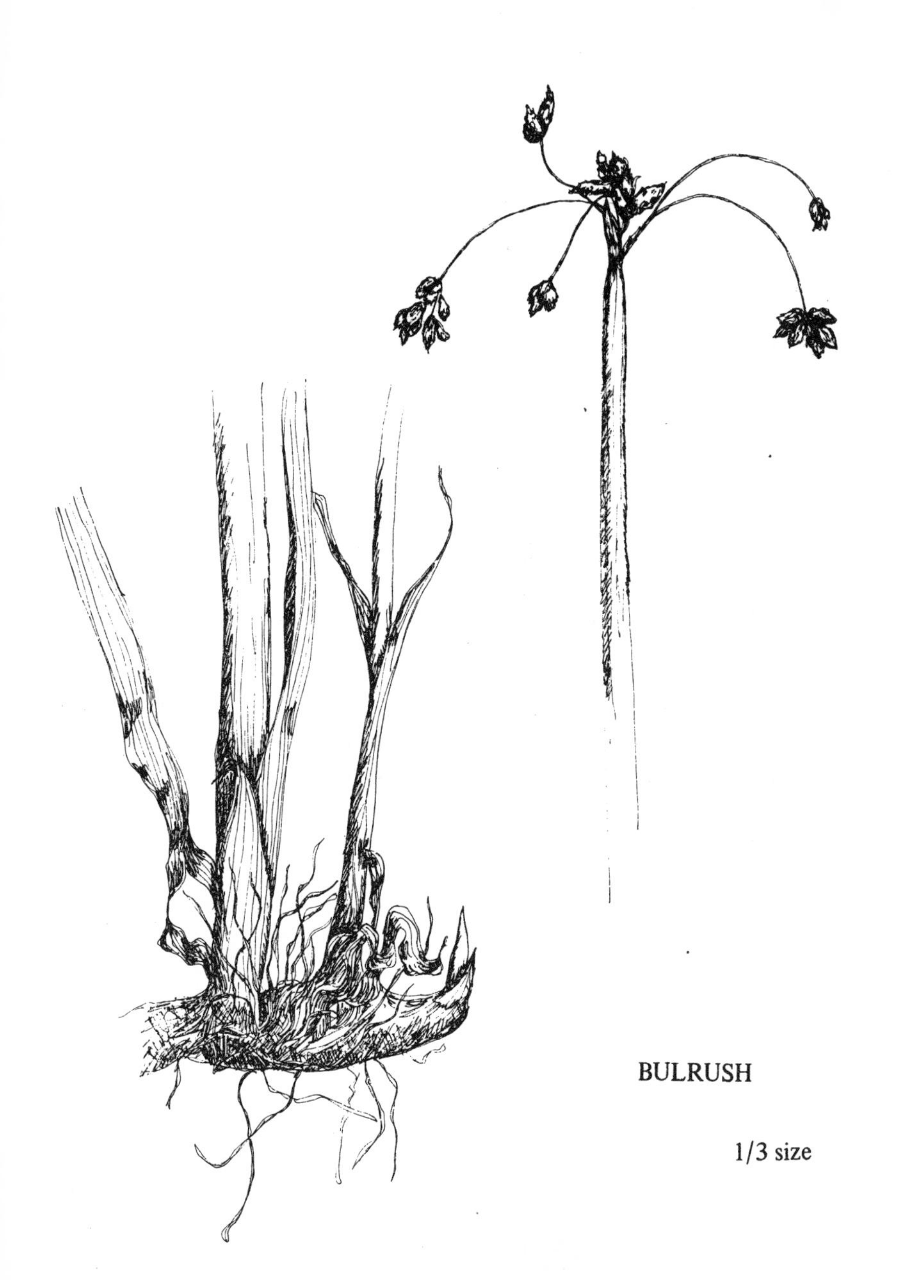

BULRUSH

1/3 size

INDIAN RHUBARB

In the early spring, you may see a weird, almost threatening-looking, pinkish stalk of flowers protruding from a running stream. Later, the flower stalk withers and the enormous leaves of Indian rhubarb *(Peltiphyllum peltatum)* begin to develop. This plant grows along the banks of streams as well as in them in the mixed-evergreen forest below 6000 feet.

I found myself at first somewhat hesitant to touch the thick stems with their bristly-looking hairs. But these hairs turn out to be rubbery rather than bristly.

The young shoots can be peeled quite readily with the fingernails and eaten raw. The taste and texture are not bad, somewhat like cucumbers or celery. However, we are not particularly fond of it.

The older shoots we peeled and boiled for about fifteen minutes. Seasoned with butter and salt, we decided the butter and salt were the tastiest part. The flavor is not really bad, but is not very exciting either. These older shoots remained somewhat tough and stringy even after boiling.

INDIAN RHUBARB

¼ size

WATERCRESS

Watercress *(Nasturtium officinale)* immigrated from Europe some years ago – an amazing fact in view of some of the remote, wild places we have seen it growing in the Sierra. It is found in small, rather slow streams. We have found it up to 6000 feet on the east and west sides of the Sierra in extensive patches.

The leaves and stems are good to eat raw or cooked. Watercress is high in vitamins A and C. The taste is very peppery or radishy. Sometimes the flavor seems too strong and is better mixed with blander greens. When boiling watercress, you can subdue the flavor somewhat by changing the water once. The full flavor is actually very good, though, once you get used to it. Watercress is good in salads and sandwiches (in small doses).

Choose tender, unblemished-looking plants. For best flavor, pick only the upper parts of each stem above the system of white roots.

WATERCRESS

life size

ALPINE SHEEP SORREL

Once you learn to recognize the leaves of alpine sheep sorrel *(Rumex paucifolius)* you will nearly always have a tasty green vegetable at hand in the high country. Also known as dock, these pretty little plants are very common in the Sierra, especially at higher elevations. They grow in a variety of habitats – in meadows, under trees, and in rocky places. Sorrel ranges between 5000 and 12,000 feet in the Sierra, but we see them most commonly in lodgepole-pine forests between 9000 and 11,000 feet. We sometimes see alpine sheep sorrel growing right near mountain sorrel (see next chapter), but the mountain sorrel clings to rock crevices while the alpine sheep sorrel chooses the flatter ground.

Like nearly all other vegetation at high elevations, alpine sheep sorrel plants are relatively small. A typical leafblade is three or four inches long. Most of the leaves are attached to the plant at ground level. The leaves tend to become partly red as summer progresses. Look for plants with the upright stalk of tiny light-green flowers tinged with red. Once you have identified the plant, it is not hard to recognize the leaves, even when no flowers are present. If you have any doubt, try a small bite of leaf. Does it have a tangy, lemony flavor?

The sour taste comes from oxalates, which in very large quantities have caused poisoning to livestock. You shouldn't worry about eating sorrel in normal amounts, though. After all, spinach and beets contain oxalates, too.

ALPINE SHEEP SORREL

life size

The Department of Agriculture lists a related species of *Rumex* as impressively high in vitamins C and A. I would be surprised if our species did not have a high vitamin content also.

If you cook the leaves as you would spinach, the result will taste something like Swiss chard with a dozen lemons squeezed on top. Too sour? We recommend changing the water once during the cooking. The leaves change color to a yellow-green when boiled. Overcooking destroys the texture, and you don't want to lose too much of that lemon flavor. We like it best boiled for one or two minutes, the water changed, and then boiled one or two minutes more. The taste is still quite sour, but only desirably so. Butter and salt complete this fresh back-country dish.

The leaves are also delicious to nibble raw. On a long, dry hike, it can be infinitely refreshing to pick a leaf or two of alpine sheep sorrel and munch as you go. They are a perfect addition to a salad of blander greens. In a sandwich they do the job of lettuce, celery, and relish.

Although these greens are quite common, we try to find an area where they are especially abundant and pluck only one or two leaves from any single plant. A substantial amount can be gathered in a short time this way without seriously hurting any one plant.

Non-native species of sheep sorrel *(R. angiocarpus* and *R. Acetosella)* have arrowhead-shaped leaves. They also taste sour but are not as succulent as our native alpine sheep sorrel. Originally from Europe, non-native sheep sorrel seems to be well established in the Sierra up to 7000 feet at least. We have found it growing miles from any road.

Large, weedy-looking species of *Rumex* without the sour taste grow at lower elevations in the Sierra, especially along roadsides and in other disrupted areas. They can be recognized by the characteristic stalk of small, inconspicuous flowers. Their leaves can also be eaten, although we have found none as exciting as alpine sheep sorrel.

MOUNTAIN SORREL

One of the best plants to know if you climb in the high Sierra is the mountain sorrel *(Oxyria digyna)*. These attractive little plants grow in rocky crevices and cracks in cliffs between 8000 and 13,000 feet. They tend to grow on the north-facing or shady side of rocks. They are very widespread near timberline and above in the Sierra. The leaves are a rounded heart-shape, about one or two inches across. The very small flowers grow on a stalk. They are greenish with a red tinge.

People often confuse mountain sorrel with miner's lettuce. The appearance is only superficially similar, however, and mountain sorrel has much tastier leaves. The leaves have the sour taste of oxylic acid, much like the alpine sheep sorrel. Mountain sorrel has more tender leaves, though, and we consider it the best fresh salad green of the Sierra. The leaves can be used in any number of ways, such as mixed in salads or sandwiches or cooked as a vegetable. But by far the best use of this plant is to pick and eat the leaves fresh as you climb one of the high mountains of the Sierra.

MOUNTAIN SORREL

life size

MINER'S LETTUCE

Miner's lettuce *(Montia perfoliata)* is one of the best known of wild food plants. It is easily recognized by the round leaves which encircle the stem just below the flower cluster. Actually, this disk consists of two united leaves. The other leaves are elongated. Miner's lettuce grows profusely in moist, shaded areas mostly below 5000 feet in the Sierra.

We have seen some people mistakenly identify jewel flower *(Streptanthus tortuosus)* as miner's lettuce. However, one attempt at a bite of this shiny, plasticky, tough leaf revealed their error.

Miner's lettuce was a popular salad with the Indians. The gold-rush miners, who had little other fresh food, used it to stave off scurvy. Miner's lettuce leaves and stems are very tender and succulent and are refreshing to eat raw. We especially like the pink-colored leaves we sometimes find under shady rock overhangs. Miner's lettuce makes a good vegetable when boiled for a few minutes and eaten with salt and butter.

MINER'S LETTUCE

life size

MONKEYFLOWERS

Some of the most pleasing wildflower gardens in the Sierra consist of monkeyflowers *(Mimulus* species) on the banks of small streams. There are many species of monkeyflowers. In the Sierra you will notice scarlet ones, pink ones, and butter-yellow ones. The most noticeable characteristic of the flower is its two-lipped structure. Three petals form the upper lip, while the two lower petals are joined to form the other lip. Often the inside of the lower lip is speckled with bright dots.

The young leaves of monkeyflowers can be eaten either raw or cooked. They soon develop a bitter taste, however, and it is hard to find them young enough. Our favorite species for this reason is the meadow monkeyflower *(M. primuloides)*. This is a dwarf variety with yellow flowers. It forms a soft mat in meadows up to 11,000 feet or so. The small leaves are much less bitter than those of the large species. Because of their small size, it is ridiculous to think of collecting any great quantity of these greens. At high elevations the leaves tend to have soft hairs which hold the dew. This does not hurt the texture for eating them, however.

MEADOW MONKEYFLOWER
life size
PINK MONKEYFLOWER
3/4 size

COLUMBINE

The columbine is well known for the intricate beauty of its flowers. If you do not know this flower, you will easily recognize it the first time you see it. The most striking feature is the way the five petals form backward-pointing spurs.

The red columbine *(Aquilegia formosa)* has bright red flowers with yellow centers. It grows in moist places, often near streams, up to about 9000 feet on both sides of the Sierra.

The alpine columbine *(A. pubescens)* has white to yellow flowers. Occasionally, we see them with a blue, violet, or pink tint. These plants grow between rocks in talus slopes and on rock ledges and similar places from 9000 to 12,000 feet.

The very young leaves are good as a salad green or may be cooked as a vegetable. As they grow older they become decidedly bitter.

Identifying columbine leaves before the flowers appear can be a problem. The best way to learn to identify them, of course, is to look at them carefully when the flowers are present. Then keep an eye out for new leaves next April, May or June.

The plant that you don't want to confuse columbine leaves with is meadow rue *(Thalictrum Fendleri).* The leaves are confusingly similar and meadow rue grows in the same damp places as red columbine. In fact, we often see the two species side by side. Meadow rue tastes exceedingly bitter, and some literature lists it as poisonous. The flowers of meadow rue are

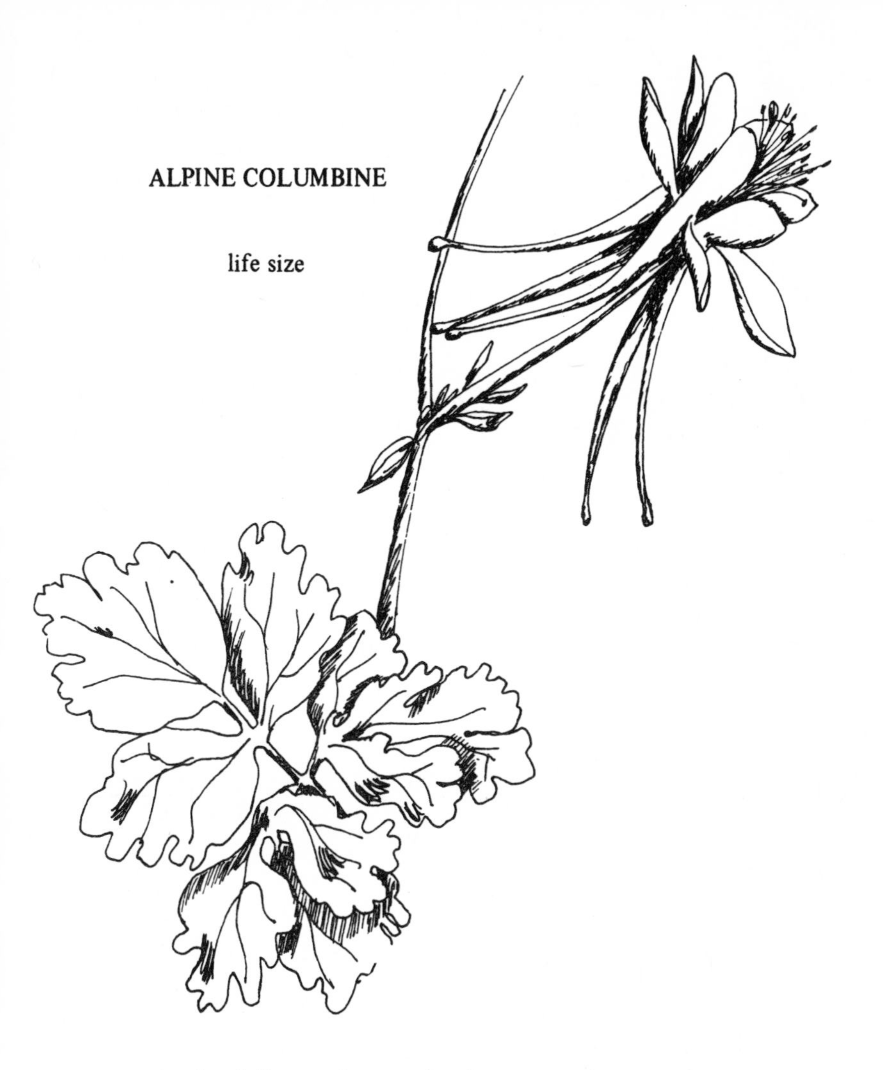

completely different from columbines. Meadow-rue flowers are little greenish things with no petals. Columbine leaflets are deeply cleft to about the center, while meadow rue leaflets are only shallowly cleft.

SHOOTING STAR

One of the most striking mountain wildflowers is the shooting star *(Dodecatheon* species). There are a number of species in the Sierra, but their appearance is very similar. The pinkish-purple to lavender petals are bent backwards giving the flower its shooting-star look. There is a bright yellow ring around the base of the petals.

The Sierra shooting star *(D. Jeffreyi)* is a common species that grows to a foot or more in height. It can be found up to about 10,000 feet in the Sierra. It grows in damp meadows, along stream banks, and in other moist places.

We tried eating the leaves of Sierra shooting star raw but decided that their texture made them unappealing to chew on. At least they do not seem to become bitter, even after the flowers are blooming. When boiled for about fifteen minutes and seasoned with butter and salt, they make a satisfactory but bland green vegetable.

The alpine shooting star *(D. Alpinum)* is a smaller species which grows up to about 12,000 feet in the Sierra. We tried its leaves with very similar results.

I might make mention here of two species which you don't want to eat. Larkspur *(Delphinium* species) and monkshood *(Aconitum columbianum)* are both deadly poisonous. The only similarity these species share with shooting stars, however, is the beautiful purple color of the flowers.

SIERRA SHOOTING STAR

2/3 size

VIOLET

Violet leaves make a tender, fresh green which can be eaten either raw or cooked. The flowers are also said to be edible, but since these are such pretty wildflowers we prefer to leave them and just pick one or two leaves from each plant. These leaves do not become bitter after the flowers develop, as many wild greens do.

There are about a dozen species of violets in the Sierra. The flowers range from white to yellow to purple. The petals of some species have colored veins. Violets have five petals; the lower one is spurred.

The species we most often encounter in the high country is *Viola adunca.* This is a small plant with violet-colored petals. The leaves are roundish to heart-shaped. We find it growing, often inconspicuously, in damp meadows among other low-growing plants. It commonly grows along grassy river banks and under trees at the edges of wet meadows. It is found from about 5000 to 11,500 feet in the Sierra. At higher altitudes, the plants are smaller. The leaves of this violet are so small that it is hard to consider them for anything more than a refreshing raw green snack.

VIOLET

life size

FIREWEED

An interesting and very common wild food plant of the Sierra is the fireweed *(Epilobium angustifolium).* The bright rose-colored flowers are very conspicuous when in full bloom. Fireweed is quick to spring up in disturbed areas, such as burned places, around old mines, and in old roadbeds. Large, dense patches of fireweed can be found in such areas. We also see it growing in apparently undisturbed areas, particularly near streams, around the edges of rocks and logs and in cracks in cliffs. In the Sierra, fireweed is seen up to about 11,000 feet. The stems are typically about three feet tall.

The very young shoots of fireweed have frequently been described as edible when cooked like asparagus. But we haven't had much luck with them. The youngest shoots we can find are thin and already tough and stringy.

The thick, mature stalks of fireweed can be peeled and the soft inner core eaten raw. We find this to be a very refreshing treat indeed. They can be peeled with a knife or with the fingernails. The pith is not strongly flavored but is sweet and pleasant, having a texture something like celery.

Fireweed leaves, either fresh or dried, can be used to make a satisfactory, if mild, tea.

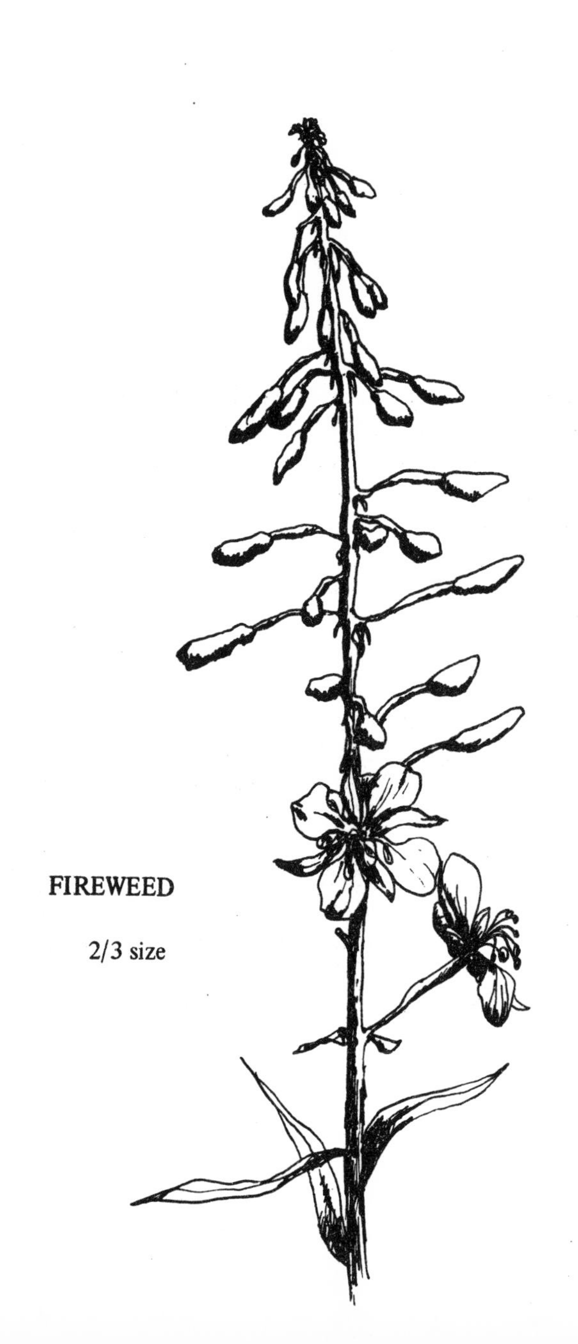

FIREWEED

2/3 size

BRACKEN

Among the first wild vegetables to appear in the spring are fiddleheads, the young shoots of the bracken fern *(Pteridium aquilinum)*. This is the most common fern of the Sierra. It grows under trees up to about 8000 feet. We have seen fiddleheads in late April and early May at 4000 feet and in July at 7000 feet. Dried-up brackens of the year before often help you in locating fiddleheads.

Pick the tender, curled-over sprouts from one to six inches high. Pick only scattered ones from large patches since each one you pick eliminates a would-be fern. Wash off the wooley hairs and boil the fiddleheads for about half an hour. Serve them as you would asparagus, with salt and butter. In our opinion, the flavor of fiddleheads is very good, but the texture is too gluey.

Older bracken ferns have caused poisoning to livestock, but the young fiddleheads are safe. Someone who is totally unfamiliar with the look of ferns could confuse a poison hemlock *(Conium maculatum)* with an adult bracken . Refer to the Yampa section for an illustration of poison hemlock.

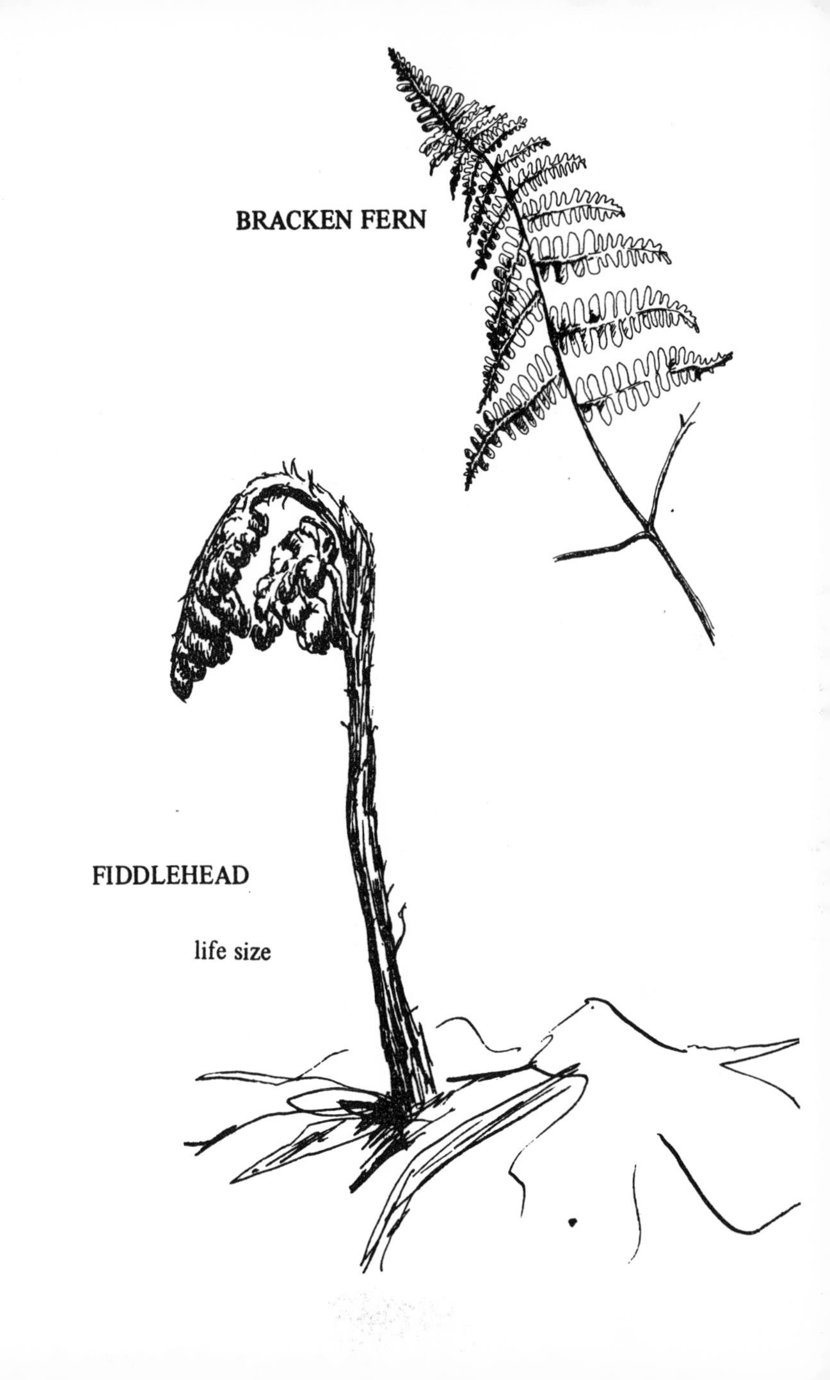
BRACKEN FERN
FIDDLEHEAD
life size

SHOWY MILKWEED

The showy milkweed *(Asclepias speciosa),* that common roadside weed, offers some of the very best vegetables to be found growing wild in the Sierra. You will see it growing in meadows and dry, gravelly and rocky places on both sides of the Sierra up to about 6000 feet. It is especially fond of open, disrupted areas such as roadbanks.

The flowers are light rose-purple and grow in clusters. Funny little hooded appendages can be seen on the inside of the petals. The plants typically grow to about three feet tall. If a stem is cut it will exude the white, milky juice which gives the plants their name.

Our first acquaintance with uses of milkweed was not as a food but as a medicine. One June in the Sierra I happened to have a small wart on my elbow. Having read of the Yosemite Indians using the juice of milkweed to remove warts, I decided to test this cure. Everyday I broke off a small milkweed stem and pressed the exuding milk into the wart. The experiment was a success. The wart was gone in about six days.

Impressed as I was with the potential of native medicine, and happy as I was to be rid of the wart , the experiment left me somewhat reluctant to try milkweed as a food. To the further detriment of my enthusiasm, milkweed is listed as poisonous to livestock.

But the season was just right to collect the unopened flower buds, and we decided to give them a try. These greenish-white bud clusters might be compared to broccoli in appearance. We

SHOWY MILKWEED

2/3 size

boiled them in three changes of water to wash away the bitter sap. We could hardly believe how delicious the cooked vegetable was. It still reminded us of broccoli tips, but the flavor was different. The closest taste we could compare it to would be peas, but that is not a very good comparison.

The first and second boilings should last about one minute each. The third boiling should be about ten minutes or until tender. We have not found any noticeable difference with this species whether we add it before or after the water boils. Season with salt and butter.

Somewhat later in the summer, usually late July to early August, the developing seed pods can be eaten. They should be cooked in the same manner, changing the water twice. The young pods should be collected when they are about one inch long. In appearance, they are similar to okra, but not in flavor. By September or October, these pods are about five inches long and filled with downy seeds.

Early in the season, about April or May, the very young shoots can be collected and cooked as above. Select shoots less than eight inches high or the tender tips of higher shoots. In this stage, the milkweed might be compared to asparagus, but again it has a flavor all its own. Before you can collect the plants in such an early stage, you must know what they look like without the flowers or leaves. It helps if you know an area from the previous year where milkweeds grow. Also, in the spring you can often find the dried stalks and pods of last year's crop. These indicate where new shoots may appear.

We sometimes find it hard to think of milkweed as a mountain plant. But it does grow commonly at reasonably high elevations and provides some of the best wild vegetables you will find in the Sierra.

MOUNTAIN PENNYROYAL

Among the best teas that can be brewed, either wild or domestic, is the mountain pennyroyal *(Monardella odoratissima).* This plant can readily be identified by its minty scent as you pass it in the high country. The flowers vary from white to purple and grow in heads at the top of the stems, which are about eight to twelve inches high. Several stems come out of each clump. You can find them growing on sunny slopes and ledges up to 12,000 feet in the Sierra.

A single stem, with the flower head still on, can be used exactly like a tea bag. It is just right for one or two cups. There is no need to boil this herb. Either fresh or dried pennyroyal will do. The tea comes out clear so you have to determine the proper strength by time or smell. The flavor of the tea can vary depending on the patch of pennyroyal you choose. Therefore, as you hike past various patches, smell them carefully to select the one you like best.

The western pennyroyal *(Monardella lanceolata)* is another species that makes a very good tea. It is found at lower elevations and tends to be taller and more branched.

The leaves of wild mints *(Mentha* species) can also be used to make tea. *Mentha arvensis,* which is possibly our only native mint, grows below 7500 feet in the Sierra. The flowers on this species all grow at the base of the leaves. Other mints, naturalized from Europe, have flower spikes at the top of the plant. These introduced species are seen in thick patches in moist meadows below 5000 feet in the Sierra. Mints can be

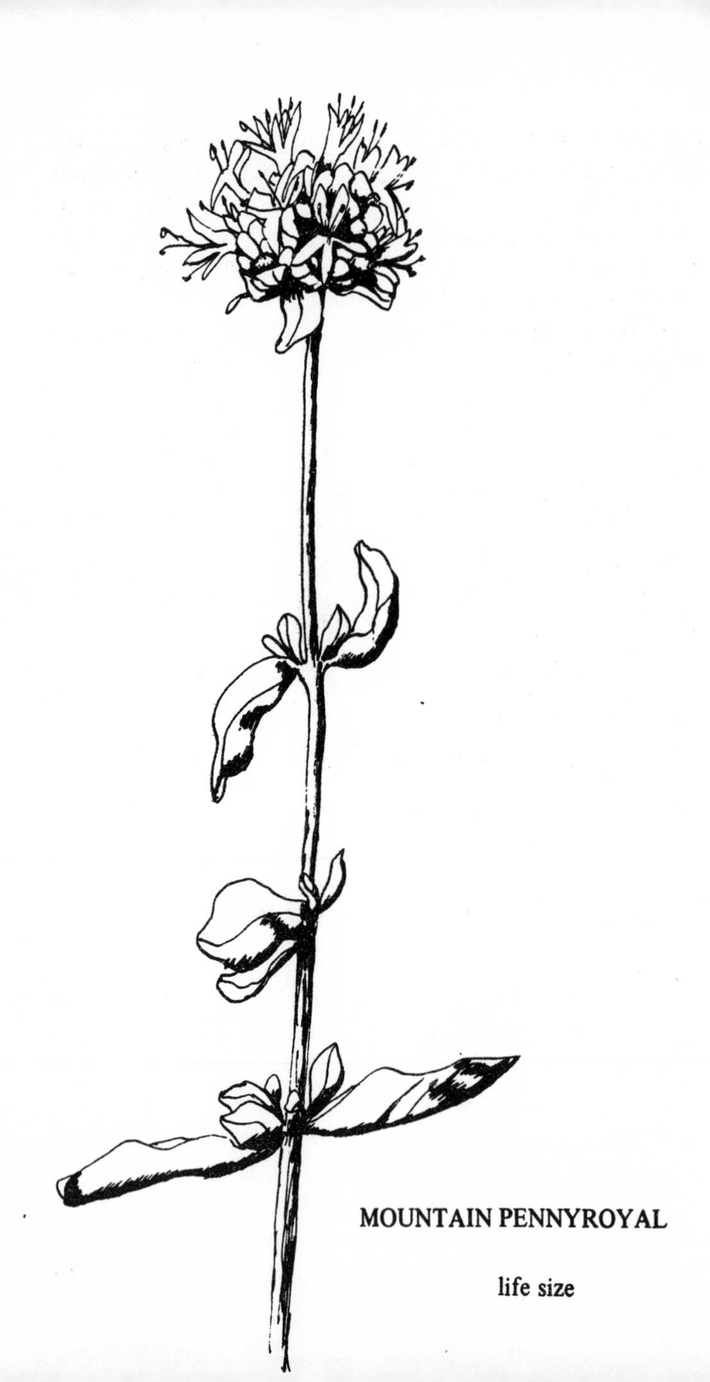

MOUNTAIN PENNYROYAL

life size

recognized by their smell. The quality of these plants is even more variable than is the case with pennyroyal. Some plants have a clear mint odor while others might be described as musky. Therefore, you should be especially selective in which leaves you choose. In addition to tea, mint leaves can be used for jellies, sauces, and as herbs in cooking.

By far our favorite "mint tea," however, is the mountain pennyroyal.

WILD GINGER

An intriguing plant of the deep, dark forest floor is wild ginger *(Asarum Hartwegii).* The prominently veined leaves are heart-shaped, about four inches across. You might overlook the flower at first. It is a strange hairy brownish thing under the leaves. Wild ginger grows in the mixed-coniferous and red fir forests between 2500 and 7000 feet in shady damp places.

The root of wild ginger can be used as a mild substitute for commercial ginger. Mary used wild ginger to flavor some pear jam she was making with fruit from an abandoned Sierra orchard. She tied the dried, chopped root in cheesecloth and boiled it with the pears. The result was very good indeed.

However, wild ginger is such a uniquely appealing part of the landscape where it grows that you certainly wouldn't want to dig up many of these roots.

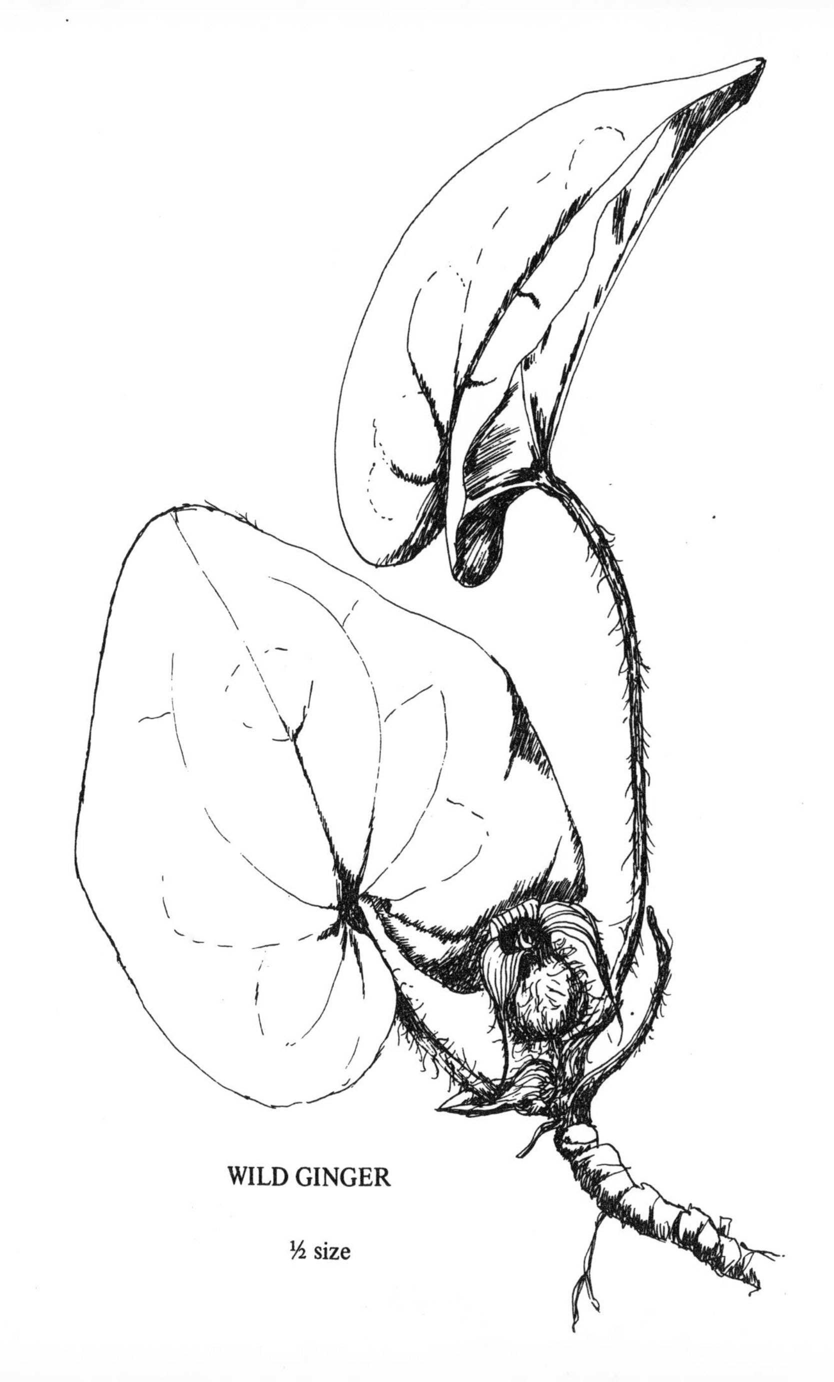

WILD GINGER

½ size

KNOTWEED

The white flower spikes of knotweed or bistort *(Polygonum bistortoides)* are commonly seen covering meadowy areas between 5000 and 10,000 feet. The slender stems give no indication of the thick rootstocks which lie horizontally just under the ground.

The very young leaves of knotweed can be boiled for greens. But the swollen, starchy rootstock was the part that made this plant an important one to the Indians. We tried these, using the edge of a knifeblade to scrape away the rootlets and outer skin. The inner portion varies from whitish to pinkish in color. Some of these roots seem all right to nibble raw, but others were definitely bitter and very tough. Boiling for about twenty minutes, however, even without changing the water, rendered the roots quite palatable. There was no trace of bitterness left, and they had become tender while still firm. When we then sliced and fried this boiled product, it became actually very good.

Knotweed is a fairly inconspicuous plant which should impress you with the potential abundance of the food it produces. Needless to say, if you are interested in digging up any plant such as this along with its roots, do so only to a very few plants from scattered localities where they grow in great abundance. Also, a digging stick and your fingers are preferable to any kind of spade or garden tool because these tools disrupt the surrounding soil and plants.

KNOTWEED

2/3 size

CAMAS

Among the most important vegetable foods for Indians all over the West were the bulbs of camas *(Camassia Quamash* and *C. Leichtlinii).* Both species of camas are found in the Sierra. The flowers are bright blue to blue-violet. *C. Quamash* is distinguished by one petal curving backward.

Camas grows in wet meadows and swampy areas between about 4000 and 8000 feet in the Sierra. Not all wet meadows in this range have camas, but where they do grow they are likely to be abundant.

The bulbs tend to be about six inches under ground, usually among a tangle of other roots, and can be rather difficult to dig up. We tried them boiled and fried. Either way they were too mucilaginous for us, although the taste wasn't at all bad.

There is a grave danger of confusing a camas bulb with a death camas bulb *(Zygadenus venenosus).* Death camas bulbs are extremely poisonous, often fatal. The bulbs look quite similar and are often seen growing in the same area. One large patch of edible camas which we found growing in a marshy aspen grove at 7000 feet had many death camas growing throughout. The arrangement of blue and white flowers certainly looked attractive, anyway. When blooming, death camas has a cluster of small white flowers. When no flowers are present, it is almost impossible to tell death camas from blue camas. Therefore, if you dig a bulb to eat, be sure there is a flower attached. Even the Indians, careful as they were, had stories to tell of Indians who had died when death camas bulbs

BLUE CAMAS
½ size

were mistakenly gathered along with the edible camas bulbs.

Camas bulbs would supply a nutritious food in an emergency, but we do not recommend gathering them to eat under normal circumstances. First, there is the danger of confusing them with death camas bulbs. Then there is the fact that we were not too fond of the texture of the cooked bulbs. Also, camas provides us with exquisite wildflowers, and for that reason alone, you would want to try very few, if any.

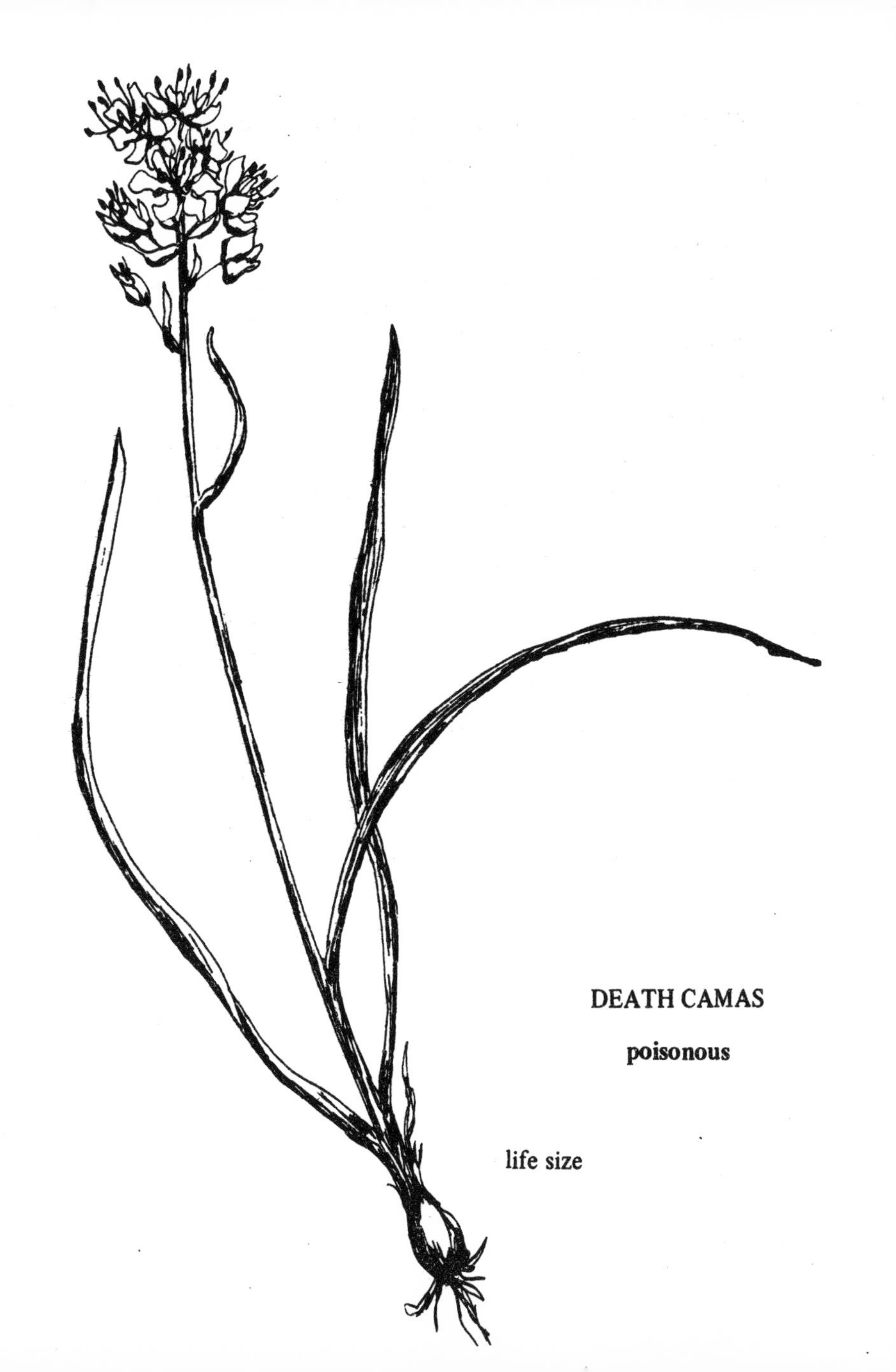
DEATH CAMAS
poisonous
life size

ONIONS AND GARLICS

The most widely known and used edible plant in the Sierra is probably the swamp onion *(Allium validum).* You might locate this onion by smell as you walk across a wet meadow or along the edge of a stream.

The flowers are purple and grow on a stalk about one or two feet high. The leaves are flat blades, often with a keel running up the center. Swamp onions can be found in dense, extensive patches in wet places throughout the Sierra up to 11,000 feet. If you find them growing in a dry-looking area, it was probably wet earlier in the season.

The bulb is not huge and round like a store onion, but rather takes the form of an elongated, thickened portion of the rootstock. It is usually found quite near the surface of the ground. In soft ground, you can just pull at the base of the onion and the bulb will come up. Otherwise, a knife stuck into the ground at an angle can be used to cut the bulb free at about the right place.

The poisonous death camas (see section on blue camas) sometimes grows alongside swamp onions. Therefore, you should be sure to have the upper plant parts connected to each bulb you collect for certain identification. Also check that it smells like an onion.

Since swamp onions grow in such thick patches, you can pick a number here and there without making a noticeable impact on the patch.

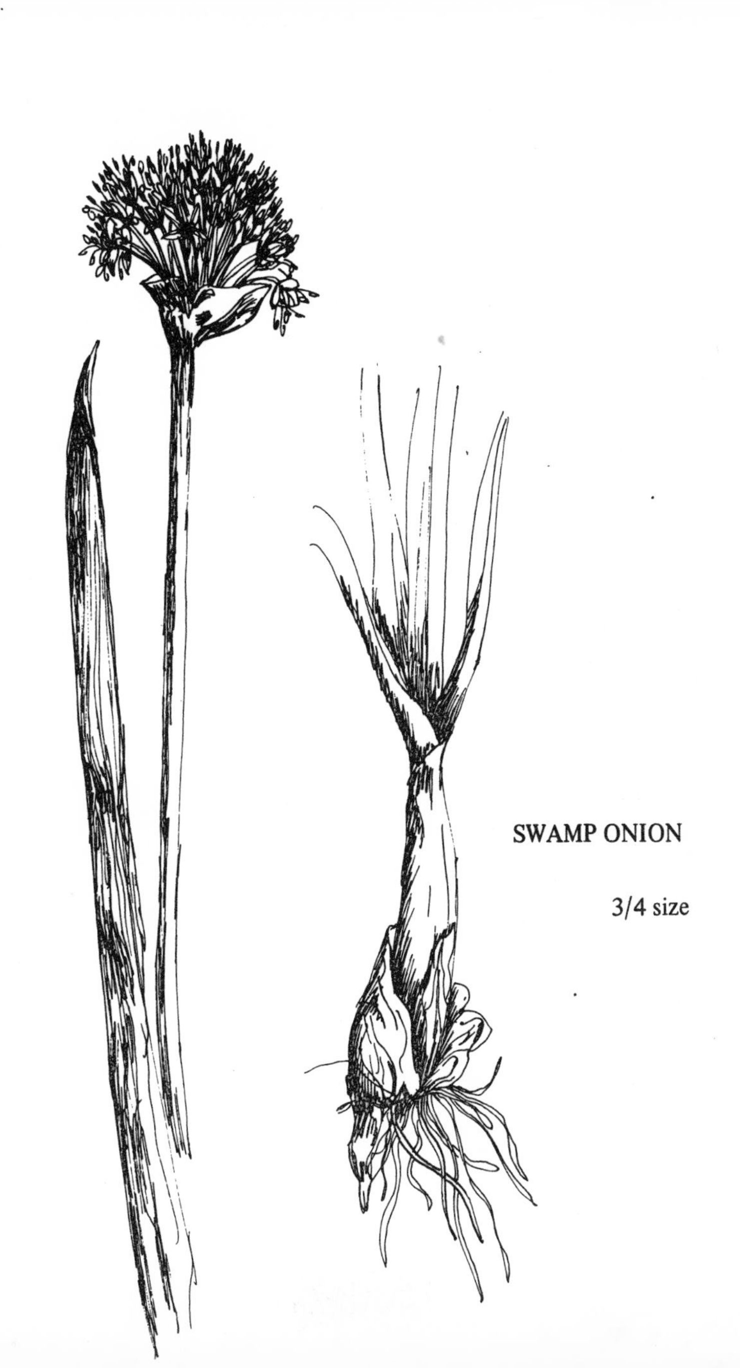
SWAMP ONION
3/4 size

The bulbs can be used any way you would use cultivated onions. In addition, the greens of young swamp onions can be used like chives. As the onion plant grows older, the upper portion of the leaves becomes tough and bitter. So, depending on its age, you may want to cut them off at some point above the bulb.

Wild onions are a nice, fresh food to liven up just about any dish in camp. Swamp onions are especially good for soup.

Backpacker's Swamp Onion Soup

Mix 3 cups water, 3 beef bouillon cubes, and about 1 cup chopped swamp onions – use greens and bulbs. Heat to boiling and simmer for a few minutes. Season with salt, pepper, soy sauce, at whim.

If you have some dehydrated meats and vegetables and some rice or bulgur, you can keep adding things until you have the whole dinner in one pot.

Sometime you will be walking through the high country when all of a sudden everything smells like garlic. Look around carefully under your feet until you find the inconspicuous flowers that are the source of the smell. The Sierra garlic *(Allium obtusum)* is a small plant. There are other species of wild garlic in the Sierra, but this one is the most garlicky and the most prolific. It is found at high elevations. We have found it as high as 11,000 feet and as low as 8000. The flower stalk is about one or two inches high and the leaf is about four inches long. In late June and July, the whitish flowers can be found in large patches on dry, sandy, sunny slopes. Usually there is just one leaf on each plant and three bracts under the flowers.

The bulbs are about a third of an inch thick. They are usually found about two inches under ground. The sandy soil makes easy digging. The best method is to use a small digging stick which you push into the ground around the bulb and pry

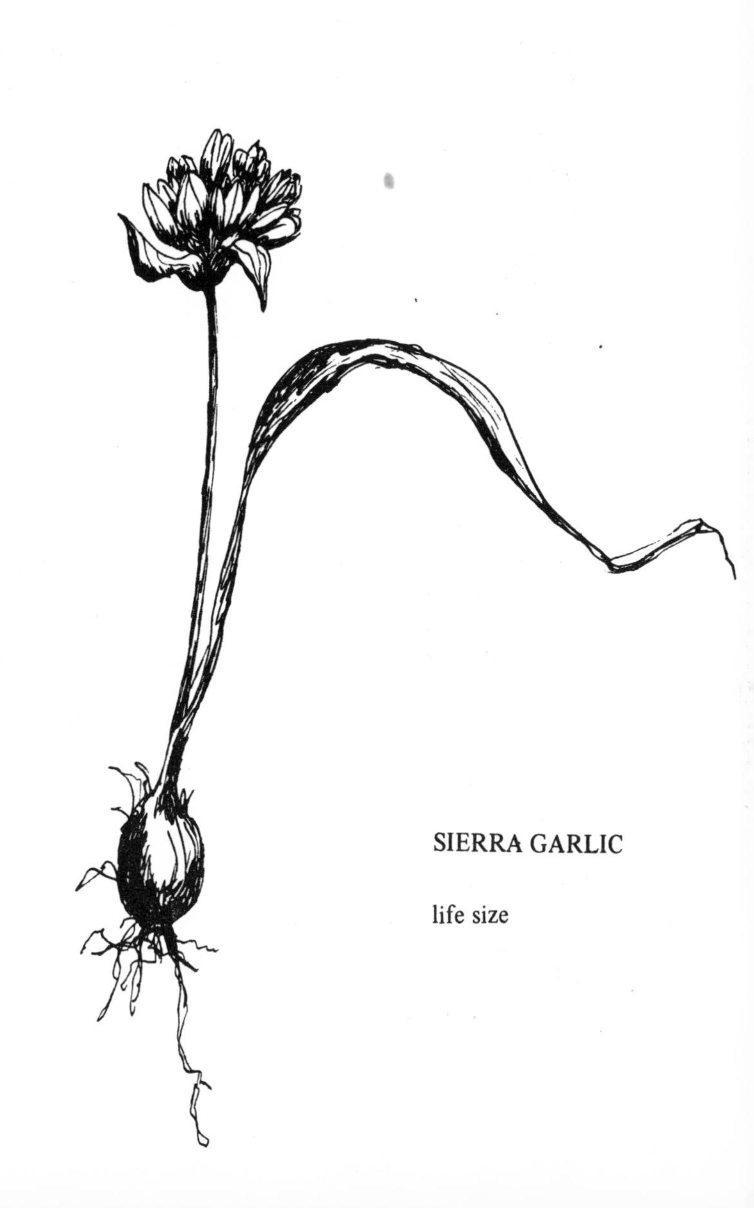

SIERRA GARLIC

life size

up. Be very careful not to break the delicate rootstalk above the bulb. You want the flowers attached to the bulb just to be sure you don't collect a death camas bulb. After the Sierra garlic flowers mature, the leaf shrivels up and the stalk breaks off at ground level, making positive identification of the bulb later in the season difficult. At least be sure that the bulb smells garlicky.

Dig the bulbs up only here and there so you do not hurt any one patch.

The little garlic bulbs are fairly mild when nibbled raw. They can be used to flavor many dishes such as soups, stews, and rice. They are also very good in omelettes. My favorite use is garlic toast. Chop up a number of garlic bulbs. Heat them with butter in a saucepan. Toast a piece of rye or French bread over the fire and pour the melted garlic butter over it.

I have read much about garlic being used as an insect repellent. An extract of domestic garlic has even been used to kill mosquitos. One day as we were collecting wild garlics while under the attack of incredible hoards of mosquitos, I decided to test the theory. While Mary applied her Cutter's, I crushed and rubbed garlic bulbs and flowers all over my face and arms. The experiment was unsuccessful, except insofar as it showed how much finer and more desirable the scent of wild garlic is than the cultivated garlic.

The Sierra has quite a number of other edible onion species. However, a word of extreme caution is in order. One species, *Allium yosemitense,* has been proposed for Threatened Species status due to its very restricted range. We feel that additional research would be likely to reveal other unique varieties that are limited to specific regions. Any collecting of wild onions should certainly be confined to the very most common species, from locations of extreme abundance.

BRODIAEA

Brodiaea bulbs were probably the most important of underground plant foods to Indians of the Sierra. These bulbs (technically called corms) are very good food, and it is impressive to see how abundant they are in the Sierra. However, brodiaeas also give us very fine wildflowers, which are lost when we dig the bulbs. Thus, you may want to know about this food, or even try it, but the days of springtime brodiaea feasts are long past – or at least no longer appropriate.

The most important brodiaea to the Indians of the Sierra was the harvest brodiaea *(Brodiaea elegans)*, as the common name implies. Each spring, the Miwoks would spend several days digging these bulbs with long digging sticks of mountain mahogany. When a great quantity had been gathered, a feast was held. The bulbs were buried in an earth oven and steamed between layers of leaves surrounded by hot rocks and live coals.

Harvest brodiaeas grow in dry, sandy and gravelly places and open forests up to about 7000 feet. The blue-purple flowers are often seen covering whole hillsides and benches. Most of them bloom in May and June in the Sierra.

The bulbs are usually found about six inches underground. The best way to dig one out is with a stick as the Indians did. A large bulb is about an inch across. It can be eaten raw but is much better cooked. It is tastier fried or roasted than boiled. It is crunchy with a flavor somewhat like a chestnut. Harvest

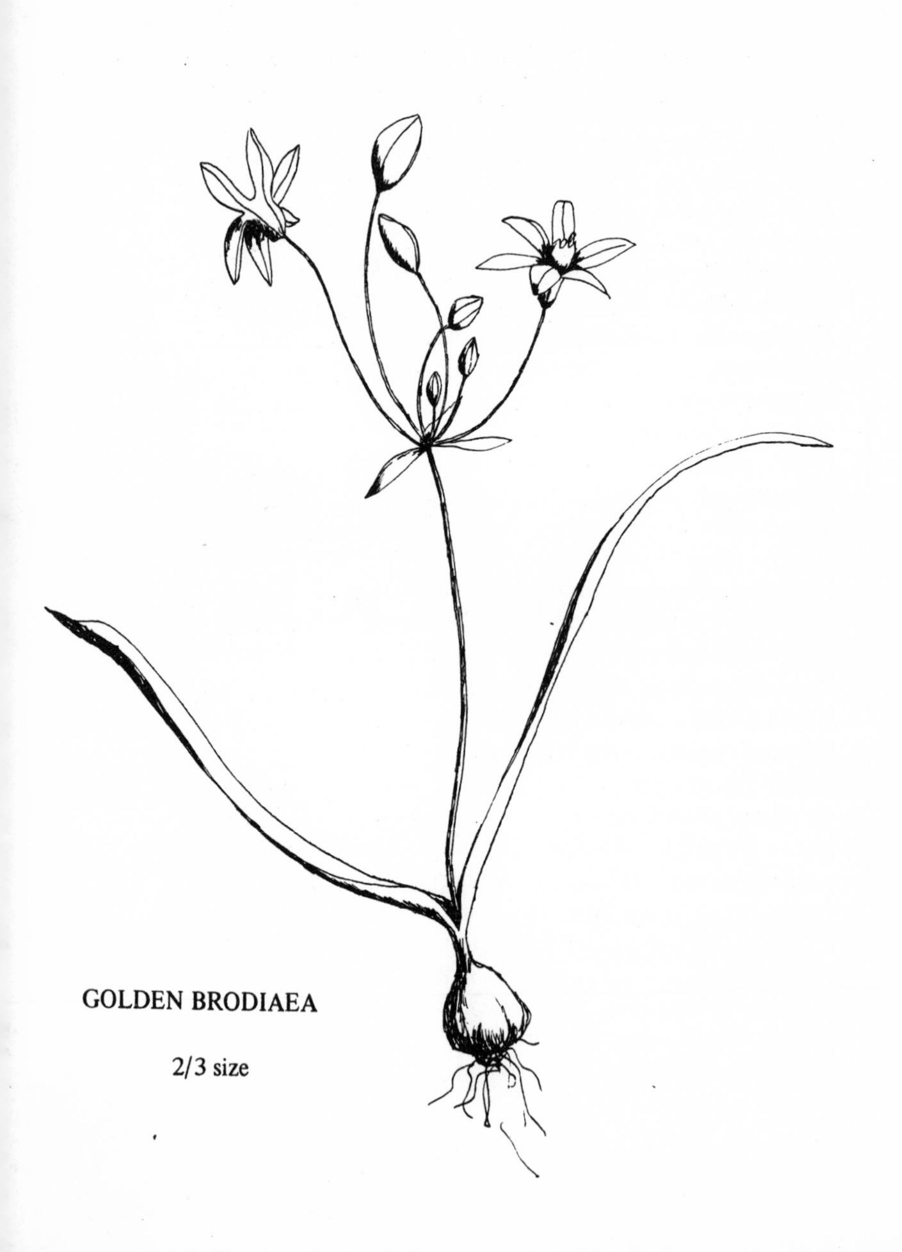

GOLDEN BRODIAEA

2/3 size

brodiaea corms are much better tasting than potatoes and hardly deserve the name "Indian potato" sometimes applied to them.

Golden brodiaeas *(B. lutea)* and a similar-looking species, *B. gracilis,* are very common in the Sierra up to about 10,000 feet. They grow in sandy places and in meadows. The flowers are golden yellow, usually blooming in June and July. *B. lutea* has dark brown or black midveins on the "petals" (sepals and petals), while *B. gracilis* has brown midveins only on the backs of the petals. The corms are smaller than those of the harvest brodiaea. The corms of either of these species can be eaten and are good, especially if not overcooked. They are not quite as tasty as harvest brodiaea bulbs, however.

We also tried the bulb of the white brodiaea *(B. hyacinthina),* which grows at lower elevations in the Sierra, below 5500 feet. The taste was similar to that of golden brodiaea.

We have seen small death camas plants bloom in mountain meadows at 9000 feet where golden brodiaeas had bloomed earlier. The brodiaea flowers appeared in late June, and the white death-camas spikes did not show themselves until mid-July. To avoid a fatal mistake, you must be absolutely sure that any bulb you try is attached to the proper plant. Also, brodiaea bulbs are apt to be wider than long, while death camas bulbs tend to be long the other way. Don't take a chance on this characteristic alone, however. See the section on camas for more information on death camas.

MARIPOSA LILIES

The mariposa lily *(Calochortus* species) has an edible bulb, but, because the flower is so precious, you would never want to dig one up. Nonetheless, you may be interested to know about such theoretical wild foods, just as we are.

The bulbs of mariposa lilies were a very important food for the Indians of the Sierra. They were collected in considerable quantity and baked like those of brodiaeas. It is probable that mariposa lilies were much more common in the Sierra that the Indians knew, when fire was a natural part of the ecology. The Indians were known to deliberately set fires to clear the meadows and the forest underbrush in order to improve conditions for certain food plants. Lightning fires were also unsuppressed before the Euroamericans arrived. Ironically, the Forest and Park Services are now working to reintroduce fire in parts of the Sierra with controlled burning.

I'll confess that curiosity once got the better of us, and we dug up a bulb next to a trail where the exquisite flower had been broken off by the boot of a careless hiker. We tried it raw, and it tasted all right, much like a brodiaea bulb. It would probably have tasted better cooked. Even digging up that bulb was somewhat regrettable, however, since the plants are perennial, which means that it might have produced another flower the year after.

Other Sierra plants which reportedly have edible bulbs include Washington lilies *(Lilium washingtonianum)* and tiger

MARIPOSA LILY

life size

lilies *(Lilium parvum* and *L. pardalinum)*. But these plants also offer us far more with their flowers than they ever could as food.

YAMPA

Yampa is a member of the carrot family, which also includes some of the most poisonous plant species. You would be wise to familiarize yourself with poison hemlock *(Conium maculatum)* and water hemlock *(Cicuta Douglasii)* so you can be certain to avoid them.

Yampa tubers were an important food to the Indians all over the west. Several states have towns, rivers and valleys which bear the name Yampa. In the Sierra, mountain meadows which look evenly white with yampa blossoms are a common sight in summer. The name yampa applies to all species of the genus *Perideridia.* Many people refer to the delicate white flowers as QueenAnne's Lace.Unfortunately, that name is also applied to several other members of the carrot family, including the deadly ones mentioned above. Other common names are squaw root and, of course, Indian potato. The Indian name yampa appeals most to us and avoids confusion.

All species of *Perideridia* have edible tubers. We have tried three species which are wide ranging in the Sierra – *P. Gairdneri, P. Parishii,* and *P. Bolanderi.* All three look very similar. Bolander's yampa has more complexly-divided leaves. Otherwise the main noticeable difference between the species is the shape of the seeds. But since all are edible, you needn't concern yourself with the final identification unless you want to (in which case refer to Munz and Keck).

We have found Yampa on both sides of the Sierra at all elevations up to about 11,000 feet. In addition to meadows,

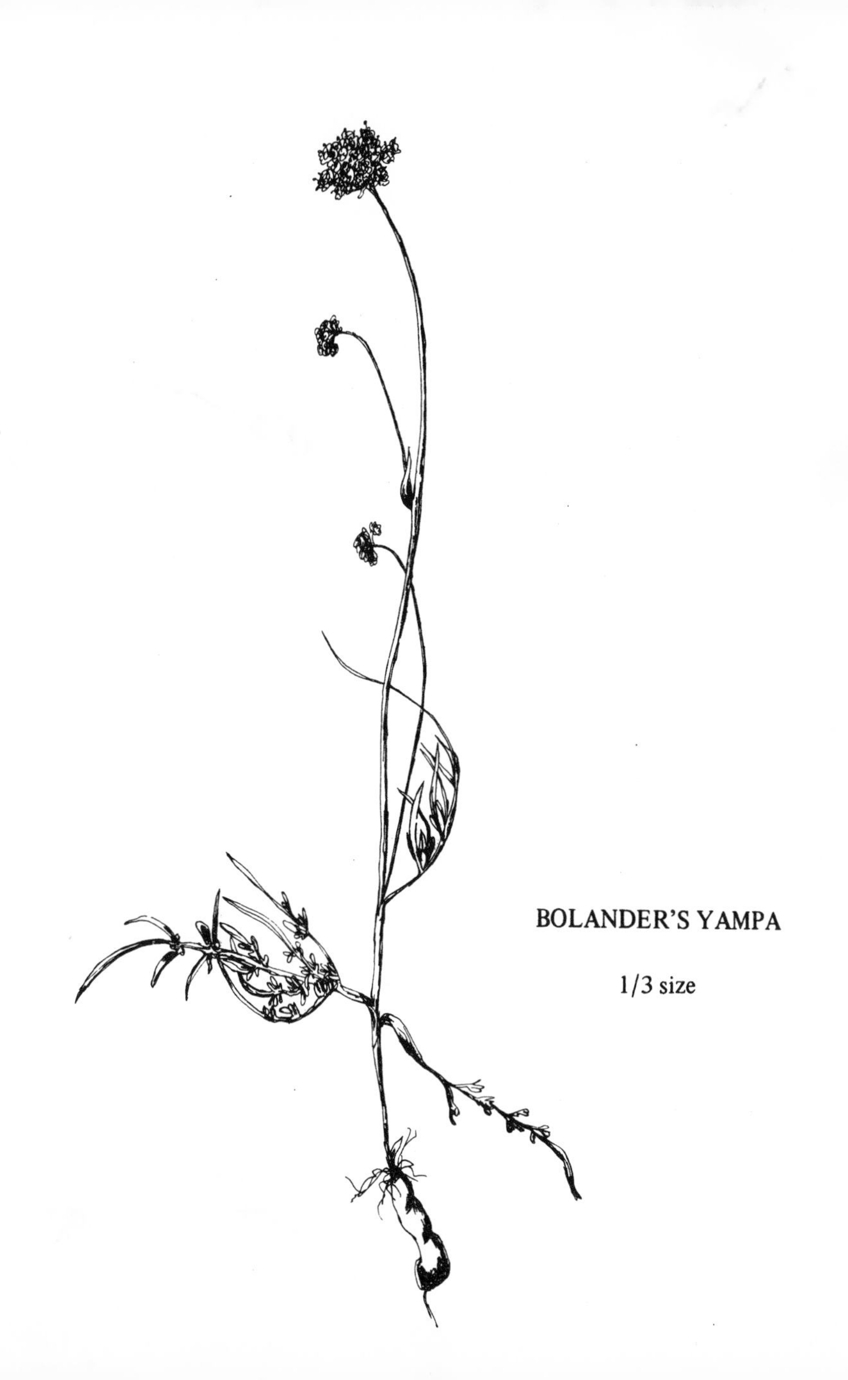

BOLANDER'S YAMPA

1/3 size

they thrive in open forests, sandy pockets, and rocky places. The white flowers can be found throughout July and August, growing incredibly profusely.

The whitish tubers are found about one to six inches below ground level. It helps if you find a place with loose dirt or sand where the digging is easy. The tubers tend to grow larger in such areas anyway. Choose a spot where your digging will not leave a scar in the surrounding vegetation. A pointed digging stick as the Indians used is ideal for the purpose. Extreme care is needed to avoid breaking the thin root right above the tuber – remember, you want the top of the plant attached for certain identification. (See Camas section for warnings about death camas bulbs.) The largest tubers we found were about two inches long and a third of an inch across.

The tubers are good to eat either raw or cooked. Our favorites are Gairdner's and Parish's yampa. Bolander's has a more spicy, radish-like taste which some might prefer, but we don't. The flavor and texture of raw yampa is something like a carrot. Cooked yampa tubers can actually be delicious. The best method of preparation is baking. Since baking is an impractical concept at the average campfire, another good style is light sautéing. The flavor is better than that of potatoes and certainly better than that of turnips. Yampa has a flavor all its own, mild and not completely describable in terms of other foods. Yampa can also be added to stews and other dishes. But the tubers are quite good enough to be eaten by themselves with butter and salt. The skins are tender, and we always eat our yampa unpeeled.

Although they grow all over the Sierra, yampa are little-noticed. For years, we shrugged the little flowers off as "just Queen Anne's Lace or something." When I sat down and really identified them, I was surprised to discover such an interesting and good-to-eat plant. We have ever since been impressed by how widespread they are in the mountains, and what a good crop of food they must have given people in times past.

WATER HEMLOCK

poisonous

½ size

POISON HEMLOCK

poisonous

½ size

SOAP PLANT

Would you pause at the idea of one plant used for fish poison, glue, soap and food? Soap plant *(Chlorogalum pomeridianum)* was used for all four by Indians in the Sierra. Also known as amole, this plant grows in dry, sandy, sunny places up to about 5000 feet. It has long basal leaves and a tall flower stalk. The flowers are white to violet with green or purple veins.

The most noticeable feature is the large bulb covered with dark-brown fibers. These hairy underground parts are often seen exposed where the ground squirrels have been digging them up.

The inner portions of the fresh bulb are slippery and can be rubbed into a good lather with water. We washed our hair with it and thought it made a fine shampoo. It is easier to rinse away than soap.

The fibers around the bulb were used by the Indians to make brushes used for hair brushes, basket scrubbers, and for sweeping acorn meal around on bedrock mortars. The brush was glued together with extract of the soap plant bulb.

The Indians could catch great numbers of fish at once by crushing the bulbs and creating a lather in a dammed-up stream. The stupified fish could be gathered from the surface. We haven't experimented with this use, and your local game warden would not like it if you did. I don't know whether there is a toxic chemical in the bulbs or if the soapy quality

SOAP PLANT
1/3 size

merely smothers the fish. Frogs were said to be unaffected. Certainly the ground squirrels don't shun the bulbs.

With some misgivings, we tried eating a soap plant bulb after drying and baking it. We let it dry for about two months and baked it at 350 degrees for 45 minutes. Surprisingly, it tasted all right, something like a potato only sweeter. However, it was too mucilaginous to really appeal to us. It was also quite fibrous, with tender portions between unchewable fibers. At least it was pretty palatable for fish poison, had good texture for glue, and wasn't at all bad tasting for soap.

MUSHROOMS – COMMENTS AND CAUTIONS

There are many edible species of mushrooms in the Sierra. We have decided to include only our four favorites. We consider each of these superior to the cultivated mushroom *(Agaricus bisporus)*. Also, each of the four is easy to identify and is not very similar to any poisonous species.

Serious mushroom hunting, of course, requires at least one book by itself. There is no good mushroom book available specifically about the Sierra, but general manuals are still quite useful. There are a number of good ones in print. We recommend *The Mushroom Hunter's Field Guide* by Alexander H. Smith, which is oriented around Michigan, *A Guide to Mushrooms and Toadstools* by Morten Lange and F. Bayard Hora, which is written in Denmark, and *The Savory Wild Mushroom* by Margaret McKenny and D. E.Stuntz,which is centered in the Northwest. These books by no means cover all the mushrooms found in the Sierra, but you should find them very helpful.

The most dangerous mushrooms belong to the genus *Amanita*. Some of these species are not uncommon in the Sierra. We have seen *Amanita muscaria,* known as the fly amanita or sacred mushroom, at various elevations up to about 10,000 feet. This mushroom is characterized by white warts on a red to yellow cap. Other poisonous Amanitas have caps which are white or shades of brown or yellow. Like the mushrooms you see in stores, Amanitas have gills under the

AMANITA MUSCARIA

poisonous

3/4 size

cap and a ring or veil around the stalk. Amanitas also have a cup around the bottom of the stem. You may have to dig around the stem to see it. You would do well to avoid any gill mushroom with a bulb or cup at the base of the stem.

Keep in mind that there are no general rules for telling a poisonous mushroom from an edible one. You simply must be certain you have identified your specimen exactly before tasting it. Don't eat too much of a new kind of mushroom on the first try. Sometimes people are allergic to a species that is generally rated edible and choice, just as some people are allergic to milk or avocados. Then there are people who say it is all a matter of whether you get along with a given mushroom or not. We heard of one man who claimed to have eaten even the most poisonous *Amanitas* by first "talking" with them and coming to terms with them. That is beyond the level of this book and its authors, however.

The main part of the mushroom plant consists of a system of underground strands called the mycelium. What we call a mushroom is just the fruiting body of the plant, a structure for distributing the spores. The microscopic spores are carried in the wind. If one lands in a suitable place – wild mushrooms are exceedingly selective -- it can establish a new mycelium network. Mycelia are everywhere in the forests and meadows. They perform the important job of decomposing organic matter into humus. They also form beneficial relationships with tree roots and may even be essential to a tree's life.

A friend from whom we learned much about mushroom hunting once told the following story. He had been walking everywhere through the woods searching for mushrooms and had not seen a one. Finally, he gave up and sat down on a log. Then he noticed some little mushrooms growing out of the log. There was another mushroom on the ground in front of him, a few over there nestled among the dead leaves, and so on. With a little experience, you will appreciate the truth of this story.

The Indians made use of other fungi besides mushrooms. A mold which grew on acorns was used to combat infection. That was a long time before antibiotics such as pennicillin from bread mold were hailed as a great discovery of twentieth-century medicine.

MORELS

One of the weirdest looking mushrooms is our favorite edible one. Morels are superb.

The morel looks something like a sculptured sponge on a stalk. The cap is thoroughly pocketed with pits separated by ridges. The spores are expelled from these pits. The only poisonous mushrooms which could conceivably be confused with morels are the brain mushrooms (certain species of *Helvella* and *Gyromitra).* There is some confusion and controversy as to whether these mushrooms are really poisonous, but it might be wisest to avoid mushrooms of this group. They can be easily distinguished from morels because their caps do not have pits. Rather, they have a convoluted surface much like an ordinary brain.

We have found morels in the Sierra at 9000 feet under lodgepole pines, at 7000 to 8000 feet under red firs, and at 5000 feet under yellow pines, incense cedar, and white fir. At 4000 feet, we found a large patch under the apple trees of an old, abandoned orchard.

The fruiting bodies appear in May and June at lower elevations and in July up higher. The ones at 9000 feet came up in late July. The best time to look for morels is after heavy spring rains.

We have eaten two species of morels in the Sierra – the common morel *(Morchella esculenta)* and the black morel *(M. angusteceps).* Both are excellent. The cap of *M. esculenta* is evenly yellowish tan in color, while on *M. angusteceps,* the

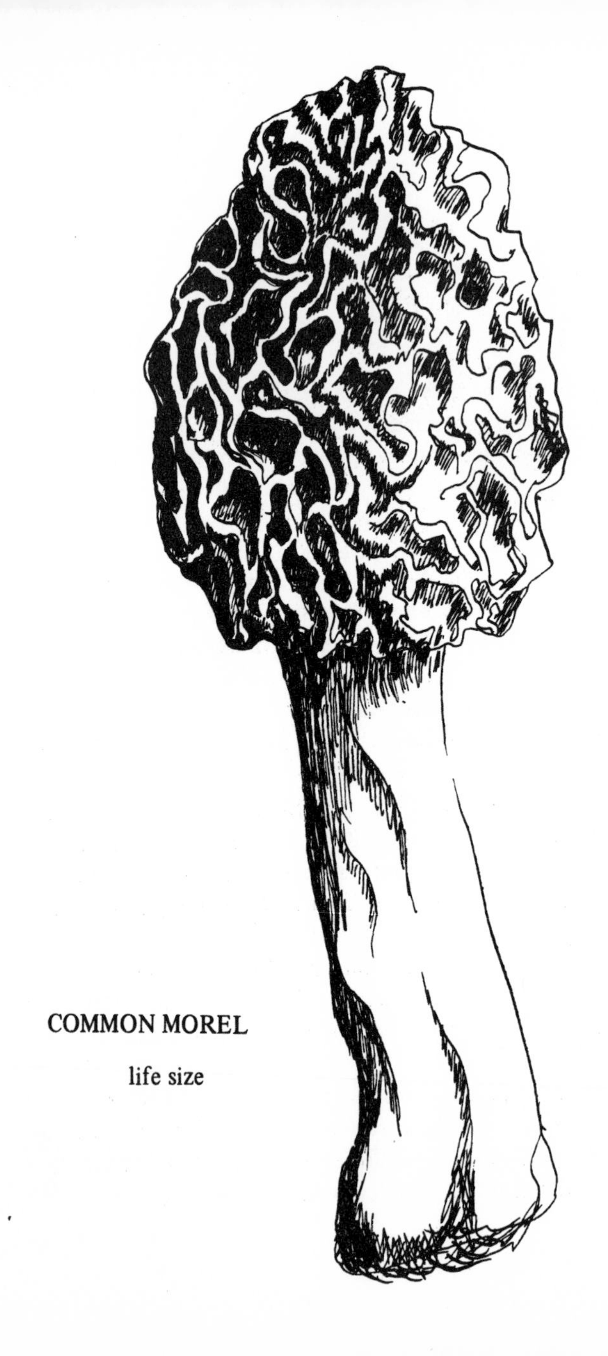

COMMON MOREL

life size

BRAIN MUSHROOM

life size

poisonous

ridges between the cups turn dark grey to black. These ridges on *M. angusteceps* form more or less continuous lines from top to bottom of the cap, while the ridges on *M. esculenta* form an irregular pattern.

Both species grow in the same habitats and at about the same time of year. One year in an apple orchard common morels came up in early May, followed by black morels in mid-May, and more common morels in late May. We have found black morels growing at higher elevations than the other variety.

Morels must always be cooked before being eaten, as raw morels have made some people sick. We have noticed a surprising lack of signs of animals eating wild morels. Perhaps that has to do with a volatile toxin.

The caps and stems of morels are hollow, suggesting that these mushrooms were made to be stuffed and baked.

Stuffed Morels

Vegetarian
- cooked brown rice or bulgur
- piñon nuts
- cheese
- salt and pepper
- green peppers
- bean sprouts

Carnivorous:
- chopped chicken livers or even hamburger
- chopped chives or swamp onions
- cooked brown rice or bulgur
- raw egg to hold it together

Stuff morels that have been washed and boiled about three minutes (discard water). Bake 15 to 20 minutes at 350 degrees.

Morels are also good sliced (boiled three minutes as usual, discarding water) and added to stews and other meat dishes. Both the cap and the stem can be used. I love to eat morels by themselves, sliced into rings, boiled and sautéed in butter.

Morels tend to come up in the same place year after year. Theoretically, you could pick every one in a patch one year and the mycelium would still fruit in the same area the next year. However, I think we should give morels every chance to establish mycelia in other areas, too. After all, this is what the fruiting bodies are for. Thus, always try to leave more than you pick in any area. Pick only young, firm ones which have not turned dark colored.

Morels are one of the most sought-after species of mushrooms. We have heard stories of French restaurants paying exorbitant sums for them. When you find a good patch of morels, you just might not want to tell anyone where it is.

CHANTARELLE

In the late fall after heavy rains, a beautiful golden-yellow mushroom appears here and there on the forest floor. The chantarelle *(Cantharelles cibarius)* grows under conifers and oaks. Where we have seen it under live oaks, we have wondered whether it is just by chance that the golden color of the chantarelle is matched perfectly by the golden color of the fallen oak leaves. The tops of chantarelles have wavy edges which often just reach the top of the leaf level on the ground. Therefore they can be hard to spot either by color or by shape. The gills, which drop the spores, are thick and blunt edged and continue smoothly from the cap down onto the stem.

Poisonous mushrooms which conceivably might be confused with chantarelles are *Clitocybe illudens* and *C. olearia.* They are orange-yellow and have thin gills descending a little way down the stem. They grow in clusters on wood, which may be buried, and are supposed to be luminescent after dark. We have never seen them in the Sierra, and they don't really look very similar to chantarelles, anyway.

Please don't pick every chantarelle you find growing in a single patch. Always leave a good number to spread their spores. In the Coast ranges, we have twice come upon a patch showing signs that someone had taken them all. Worse yet, these people had cut the tops of the chantarelles off with a knife. (They probably don't like to touch "dirt.") Thus, they wasted a good edible portion of each mushroom, while still

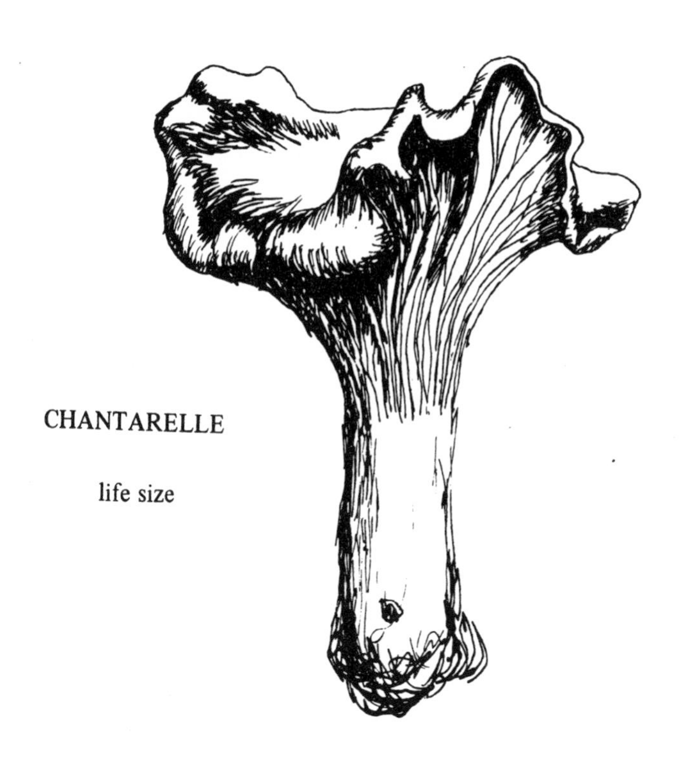

CHANTARELLE

life size

curtailing its reproductive effort. It is much better to pull them from the ground with your fingers to get the whole mushroom, all of which is edible.

When picking chantarelles, choose only young, firm ones which are free of worms.

Chantarelles have a delicious, mild taste and are good to eat by themselves, sliced and sautéed slowly in butter. Chantarelle omelettes are my favorite use of this choice mushroom. They look just as fine as they taste. In any other dish they may be used in place of (blah) commercial mushrooms.

KING BOLETE

The king bolete or cep *(Boletus edulis)* is another gourmet mushroom found in the Sierra. It can be very large, about nine inches across, and both cap and stem are edible. The stem is very swollen or bulbous in shape. The cap is somewhat sticky when fresh. It is a pale brown on top and white underneath. It does not have gills underneath like the cultivated mushroom, but instead has a spongy looking surface of pores which drop the spores.

This species normally fruits in the fall, but we have also found them during an exceptionally wet spring in the mixed coniferous forests of the western Sierra slope.

Avoid any pore mushroom which has red pore mouths under the cap or which stains blue or green when bruised, since some such species are reported to be poisonous.

King boletes are best when young. Older ones tend to be wormy. In fact, a bolete seldom reaches full size before a number of animals have eaten at it. Both the cap and stem are delicious sliced and sautéed in butter. You may want to remove the pore layer, especially with large specimens.

KING BOLETE

life size

PUFFBALLS

Puffballs are the easiest of all mushrooms to identify, and all of them are edible. The only danger is that you might mistake a deadly *Amanita* button (immature mushroom) for a puffball. Therefore, take any puffball you plan to eat and cut it in half longitudinally. If it is an amanita, you will see the outline of its gill and stem structure.

Only a young puffball which is solid feeling and evenly white throughout is good eating. Eventually, the inside turns yellowish and then dark and dries out as the spores mature. The skin becomes dry and cracks open. At this stage, but not before, you can step on it and watch the spores puff away into the wind.

The most interesting puffball in the Sierra is the Sierra puffball *(Calvatia sculpta)* – but don't pick it. This is a beautiful little creation with pyramidal ridges and cracks all over the surface. They are thought to be much less common than they used to be in the Sierra as tourists have picked them for their scenic value, as well as their food value. Although there are records of Sierra puffballs about twenty inches in diameter, we usually see specimens about three or four inches thick. We have seen them up to about 10,000 feet in meadows. They're nice to look at, but let's not pick them. There are plenty of smaller, plainer, and more common puffballs to eat in the Sierra.

Quite a variety of puffball species grow in the Sierra at all elevations. Some are smooth on the surface, some bumpy or

SIERRA PUFFBALL

life size

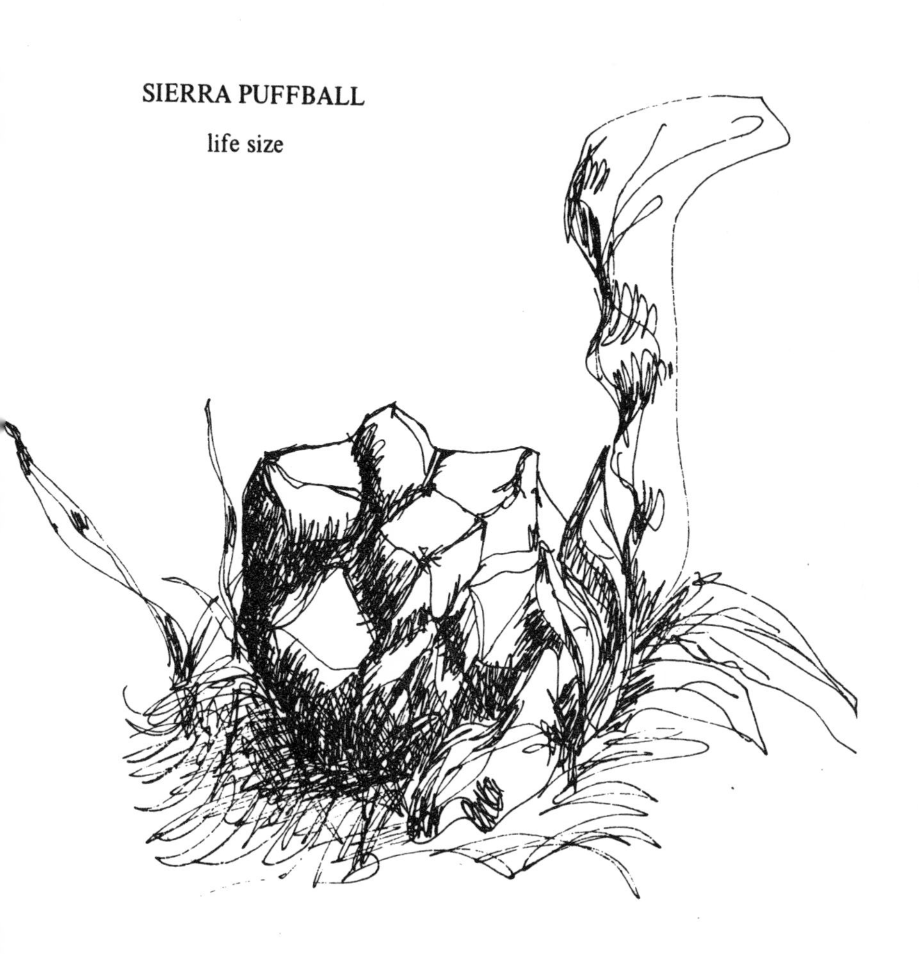

spiny. Some are roundish, some pearshaped. We have seen whole meadows dotted with puffballs about half an inch in diameter. Other common types are two or three inches in diameter and grow under conifers. Look for ones that are evenly white inside and out.

The flavor of puffballs is quite different from that of other mushrooms. It is especially suited for flavoring soup.

Fungus Soup

Cut up 1 to 2 cups of puffballs and roll in flour. Lightly sauté in 2 tablespoons of butter. Slowly stir in 3 cups milk and heat but do not scorch. Add a little chopped chives (swamp onions) or bacon crumbles if desired. Season with salt and pepper.

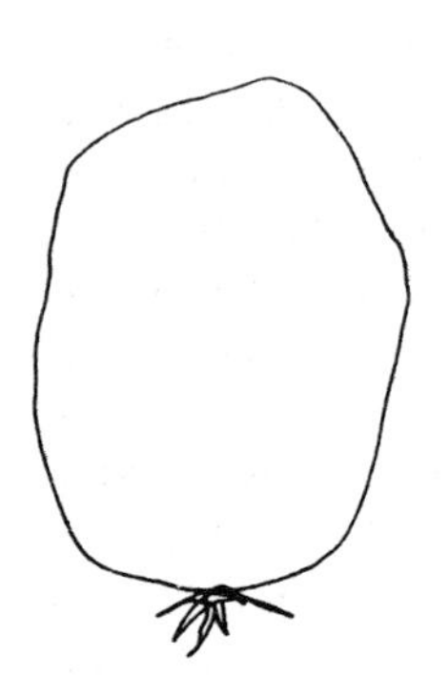

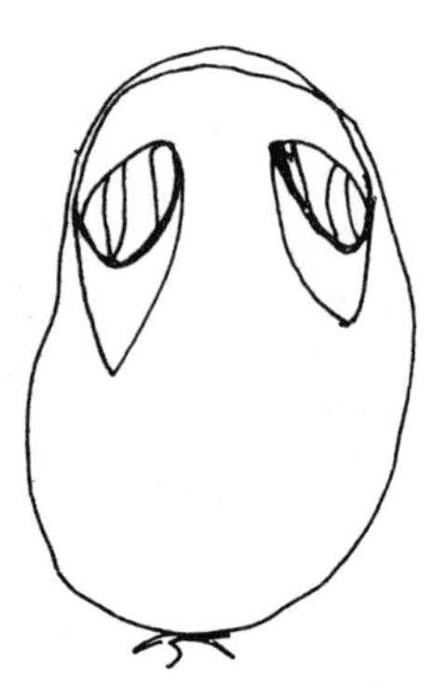

cross sections

PUFFBALL

AMANITA BUTTON

poisonous

ALLEGEDLY EDIBLE BUT UGH!

There are always pitfalls in believing everything you read or hear, but disillusionment in the case of wild foods can be a most unpalatable experience. We have had our share of such experiences, although the results have been nothing worse than a bad aftertaste. Differences of opinions and digestions probably account for some of these surprises. In other cases, mistakes have perhaps been made where academic research was not checked with personal experience.

We have never been able to stand the taste of juniper berries, the fleshy cones of juniper trees *(Juniperus species).* Yet many sources speak of Indians, pioneers, and outdoorsmen brewing a tea with them, adding them to soups, or even eating them outright. Juniper berries are, incidentally, well known for flavoring gin. But we know we couldn't eat these berries even if we were starving. One spring in the Sierra a bear helped to reinforce our opinion. We were backpacking up a trail one morning in late May, perhaps the leanest season of the year for black bears. The tracks of a bear preceded us up the trail. I suppose he had been out of hibernation for a couple of months, and there was still not much new growth of anything – this was at about 7000 feet. Apparently he had been eating nothing but juniper berries lately because they were the only solid parts of his droppings – and the droppings were spaced about one hundred feet apart all the way up the trail.

WESTERN JUNIPER

life size

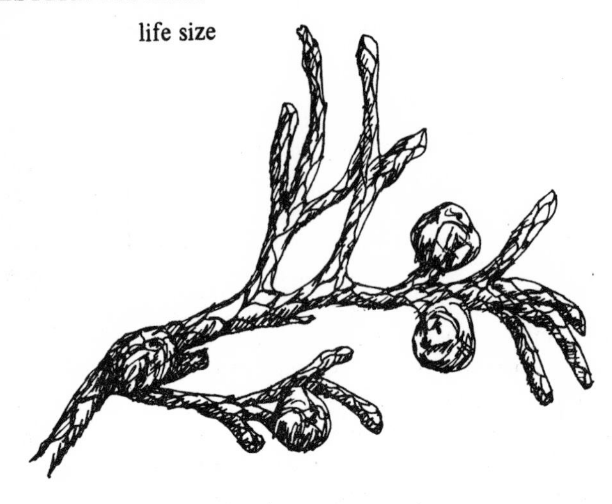

The marsh marigold *(Caltha Howellii)* is said to be poisonous raw, but the leaves are recommended to be eaten cooked. We tried boiling the leaves for about fifteen minutes and didn't like them at all. The texture was fine and they weren't bitter. We just found the taste displeasing. Buttercup *(Ranunculus* species*)* leaves are eaten by some people after a similar volatile toxin has been cooked away. However, some species of buttercups are apparently poisonous to some people even after cooking.

Strangely, the Indians seem to have been able to eat some plants that would be poisonous to us. For example, they sometimes ate poison oak *(Rhus diversiloba)* leaves with their acorn bread. Certain other plants, such as lupines *(Lupinus* species), would be poisonous to eat fresh but were eaten after suitable processing.

One anthropological study of Indians in the Sierra lists the root of the corn lily or false hellebore *(Veratrum californicum)*

MARSH MARIGOLD

life size

as having been roasted and eaten. Unfortunately, it is one of the most poisonous plants of the Sierra. The root was actually eaten by Indians when they wanted to commit suicide, or at least become sterile. Deer avoid corn lilies, but there are many records of livestock losses from this species. Conceivably, the Indians could have had some method of processing out the poison, but cooking alone would not be enough. The above-mentioned study of Indian culture is a good one, and the mistake probably resulted from a communication problem with one of their informants. The corn lily is the plant that looks like an ear of corn when it first comes up in meadows from 5000 to 11,000 feet. Eventually it becomes a rather tall plant with a stalk of white flowers.

We could go on and on with examples of allegedly - edible - but - ugh plants, but we have at least covered some of the most significant ones. It may reassure the reader to know that, apart from certain species in this chapter (such as poison oak), we have personally eaten all of the wild food plants discussed in this book. We cannot clain such intimate knowledge of the various poisonous plants mentioned throughout the book, however, for which we must rely solely on second-hand information and allegations.

USEFUL BOOKS

Abrams, Leroy, *Illustrated Flora of the Pacific States,* 4 vols., Stanford University Press, Stanford, California, 1926 - 1960.

Barrett, S. A. and E. W. Gifford, *Miwok Material Culture,* Yosemite Natural History Association, Inc., Yosemite National Park, California.

Harrington, H. D., *How to Identify Plants,* Sage Books, The Swallow Press, Inc., Chicago, 1957.

Jepson, Willis Linn, *A Manual of the Flowering Plants of California,* University of California Press, Berkeley and Los Angeles, 1925.

Kephart, Horace, *The Book of Camping and Woodcraft,* The Century Co., New York, 1909, later editions by The MacMillan Company.

Kingsbury, John M., *Poisonous Plants of the United States and Canada,* Prentice-Hall, Inc., Englewood Cliffs, N.J., 1964.

Lange, Morten and F. Bayard Hora, *A Guide to Mushrooms and Toadstools,* E. P. Dutton and Co., Inc., New York, 1963.

McKenny, Margaret and D. E. Stuntz, *The Savory Wild Mushroom,* revised edition, University of Washington Press, Seattle, 1971.

Muir, John, *The Mountains of California*, Doubleday and Company, Garden City, New York, 1961.

Munz, Philip A., *California Mountain Wildflowers*, University of California Press, Berkeley and Los Angeles, 1963.

Munz, Philip A. and David D. Keck, *A California Flora*, University of California Press, Berkeley and Los Angeles, 1959.

Nearing, Helen and Scott, *The Maple Sugar Book*, Schocken Books, New York, 1950, 1970.

Neihardt, John G., *Black Elk Speaks*, University of Nebraska Press, Lincoln, 1961.

Smith, Alexander H., *The Mushroom Hunter's Field Guide*, revised and enlarged, The University of Michigan Press, Ann Arbor, 1963.

Storer, Tracy I. and Robert L. Usinger, *Sierra Nevada Natural History*, University of California Press, Berkeley and Los Angeles, 1963.

INDEX